W9-CNC-485

Safe and Effective Use of Crop Protection Products in Developing Countries

Contents

Preface

There has been a great deal of good news about development over the last 30 years: average life expectancy at birth worldwide increased by more than a third, the infant mortality rate in developing countries fell by more than half, real per capita income there increased on average 3.5% a year, and the share of the population suffering from chronic undernutrition in these countries dropped from more than one third to about one fifth. The developing world achieved gains in the last 30 years that took the industrial world a century, and all these positive changes took place while world population increased by more than 2.5 billion. More people live longer and fuller lives than ever before in human history.

This book looks at one form of that technological progress—the use of chemical plant protection products (also known as pesticides)—that has made an important contribution to improving global food security during the last three decades. Pesticide use has both supporters and critics, as described in Chapter 1. Major substantive issues that will continue to confront policymakers and others interested in more effective crop protection will be how to balance social costs with social gains from the use of chemical pesticides—that is, how to reduce crop losses while minimizing pest resistance to pesticides and harmful side effects to human health and the environment. Although quite a lot still needs to be learned about the side effects of chemical pesticide use, the actual and potential social and economic costs from excessive pesticide use make it desirable to reduce the use of chemical pesticides without reducing needed crop protection. The health and environmental risks of the remaining usage should be minimized.

The research that led to this report was undertaken as part of the Risk Fund set up by Novartis (then Ciba-Geigy) in 1988 to support its business activities in the Third World. The fund is intended for commercially oriented projects that require especially extensive services or preparations or expensive support. It is also designed to support projects that require high initial investments. The Risk Fund supported projects in the Dyes and Chemicals, the Pharmaceuticals, and the Agricultural Divisions.

The study was sponsored by the Novartis Foundation for Sustainable Development. All references to the company and the foundation throughout this report are to Novartis for consistency's sake.

The results of this pilot project should help companies involved in the crop protection business to adapt their future marketing concepts to specific sociocultural conditions in order to make plant protection products safer to use. We also hope that others can benefit from the lessons we learned during the project, and to that end we are making all the data generated in the project available to interested parties and the public.

In the end, each society must decide what kind of risks it wants to live with. This determines the portfolio of the products and services it allows through regulation. While the decisions will always be the result of a political process and therefore open to lobbying by different interest groups, the database upon which these decisions are made must be scientifically sound. When the Food and Agriculture Organization first published its *International Code of Conduct on the Distribution and Use of Pesticides*, the international Association of the Plant Protection Industry welcomed it and promised to support all endeavours to improve the state of affairs. This publication is meant to be part of a constructive and open dialogue about the safe and effective use of plant protection products.

Acknowledgements and Contributors

This research could not have been undertaken without the commitment and hard work of scores of people over a good number of years. The Novartis Foundation for Sustainable Development would like to thank the many members of the Project Teams in the three countries studied—India, Mexico, and Zimbabwe—and all the farmers who participated in the research.

The results of the research were analysed and reviewed by the members of the International Steering Committee. Unless otherwise indicated, the following contributors can be reached c/o the Novartis Foundation for Sustainable Development, WRO-1002.11.59, Postfach, CH-4002 Basel, Switzerland, or by e-mail at <novartis.foundation@group.novartis.com>.

Beda Angehrn is President of Agrovision AG, St. Gallen/Switzerland, and can be reached at Erlackerstrasse 61, CH-9303 Wittenbach, Switzerland, or by e-mail at <BedaAngehrn@.compuserve.com>.

Jost Frei was the International Project Leader and former Head of Region Africa within Novartis Sector Crop Protection, Business International.

Al Imfeld is an independent journalist. He can be reached at Konradstrasse 23, CH-8005 Zürich/Switzerland, or by fax at (41-1) 272 17 51.

Bernd Kupferschmied is the former Head of Product Management in the Business Unit, Insecticides, at Novartis.

Klaus M. Leisinger is Executive Director and Delegate of the Board of Trustees of the Novartis Foundation for Sustainable Development and Professor of Development Sociology at the University of Basel.

Hans Pfalzer is a former specialist of the farmer support team within Novartis Crop Protection.

Hiru Punwani, formerly with Novartis in Hindustan, India, was the Project Leader for India.

Peter Rieder is Professor for Agricultural Economics at the Federal Institute of Technology (ETH), Zurich, Switzerland. He can be reached at Institut für Agrarwissenschaften, ETH-Zentrum, CH-8092 Zürich/Switzerland, or by e-mail at <peter.rieder@iaw.agrl.ethz.ch>.

Karin M. Schmitt is head of Social Development at the Novartis Foundation for Sustainable Development.

Jean-Paul Serres was the Project Leader for Mexico.

Linda Starke is an independent editor specializing in international environment issues. She can be reached at 5405 Sherier Pl., N.W., Washington DC 20016, USA, or by e-mail at <Starkeat50@cs.com>.

Hema Viswanathan is Senior Vice-President, Social and Rural Research Institute, Delhi, India. She can be reached in Mumbai, India, by fax at (91-22) 432 38 00.

Klaus von Grebmer is the former Head of Public Affairs of Sector Crop Protection at Novartis.

Andreas Weder of Novartis Iran was the Project Leader for Zimbabwe.

Montague Yudelman is the former Director of Agriculture at the World Bank and a Fellow at the World Wildlife Fund in the United States. He can be reached at 3108 Garfield St., N.W., Washington DC 20008, USA.

Executive Summary

Safe and Effective Use of Crop Protection Products in Developing Countries is the report of a seven-year-long research programme that looked at ways to achieve safer, more effective, and reduced pesticide use among low-income farmers in developing countries. The research was undertaken as part of a Risk Fund set up by Novartis in 1988 to support its business activities in the Third World; the study was sponsored by the Novartis Foundation for Sustainable Development.

An International Steering Committee was set up to oversee the project. At the outset, the project wanted to clarify:

- what factors hinder the safe and effective use of pesticides in developing countries;
- what sort of groundwork can or must manufacturers, in collaboration with other institutions, lay to eliminate these factors; and
- what, in a given social cultural context, are the communication methods best suited to the farmer community.

The selection of countries to be included was guided by a desire to carry out a social marketing experiment in three nations that represented similar economic development but widely disparate sociocultural environments as well as agricultural practices. India, Mexico, and Zimbabwe were chosen; combined, they are fairly representative of agriculture in the developing world. Novartis affiliates in each country helped set up local project organizations that took responsibility first for baseline studies in 1992–93 on farmers' knowledge, attitudes, and practices (KAP) regarding pesticides, and then for implementation of the communication campaigns that followed, starting in 1993. Control and test areas were established in each country.

Before the communication campaigns were launched, it was important to consider the availability of media for reaching small farmers as well as to identify the media that were most attractive and credible. In Mexico, radio has both audience appeal and reach, and was used to reinforce messages delivered during special farmers' meetings. In India, in contrast, films are an important source of entertainment and an escape from daily drudgery; the use of television and VCRs allowed a film to reach all test villages without spilling over into neighbouring regions. In Zimbabwe, folk theatre and the oral rendition of stories and messages is a lively, vibrant, and yet ancient tradition. It has instinctive and almost reflex appeal, and this alone made the medium the right choice for this country.

It turned out to be very difficult to assess or measure the impact of isolated interventions in a dynamic social environment where life in general changes constantly through improved communication and where competitive companies and agricultural extension services are active. The programme over such an extended period of time in such diverse cultures led to a learning process that raised additional questions and issues to be analysed, and thereby some issues that later on were found to be of importance were not covered by the initial baseline study.

The project ran into a few setbacks and unexpected obstacles during its five years, which is not surprising, given the nature of doing research in a changing world. First, in retrospect there were a few specific problems in the selection of areas for the project. In Mexico, the climates of the test and control areas were known to differ slightly, but we did not expect that the difference would vary with years or seasons. (Generally, similarity between the two areas is important, but even more important is the requirement that any difference that might exist should not vary during the intervention period.) In India, the fact that the control area surrounded the intervention area proved more of a problem than we expected in terms of hindering our ability to afford a spillover effect.

In addition, several economic and information factors that could not have been anticipated may have affected the study's results. A drought in Zimbabwe, for example, changed farmers' perceptions of the importance of effectiveness versus safety. An earthquake in the test area in Mexico in October 1995 temporarily set back efforts to interest local farmers in the intervention programme, and in the control area in Mexico, a communication programme by a competitor around the same time as the project may have led to many of the favourable changes in knowledge levels that we found there at the end of the study period.

Nevertheless, it is possible to track some specific changes in farmers' attitudes and practices when measured against four hypotheses for the project that were established by the Steering Committee.

Hypothesis: Through communication and training we will improve practice in:

- skin protection,
- preparation of spray solution,
- washing of body and work clothes,
- spraying and application, and
- maintenance of spraying equipment.

In Mexico, personal safety is the area where major improvements were detected. Persistent changes were found in the use of shirts, boots, and footwear;

in the washing of work clothes and hands; in sprayer cleaning; and in avoidance of the chili can as a measuring device. Major attitude changes were detected at focus group sessions. The project therefore had a favourable impact on comparatively simple, cheap safety practices, while more cumbersome practices did not change for long or were used even less.

In the test region in India, the practice of having a full body wash after spraying rose significantly and was sustained. Washing of work clothes worn while spraying was at a high level in 1996 and was sustained in 1997. The use of gloved hands for mixing pesticides showed a small improvement but dropped sharply after cessation of intervention. Significant improvement in the maintenance of spraying equipment was registered. Farmers' and spraymen's practice of taking precautions before breaking for food, drink, or a smoke was at a high level, and it persisted in the project area. And the use of a full-sleeved shirt and full trousers went up significantly, but after cessation of intervention these dropped somewhat, although still registering positive change.

In Zimbabwe, personal hygiene and skin protection were already at a high level when the project started. The project had a further positive impact on the use of gloves, proper shoe attire (gumboots and so on), and the use of full body protection. Another significant increase was reported in the regular washing of work clothes. Project managers also recorded a much higher percentage using gloves when mixing pesticides, which was a desired result. Overall, the project had a positive impact on reported attitudes regarding personal hygiene and skin protection.

Hypothesis: Through communication and training we will improve practice in:

- optimization of quantity of plant protection products used and of spray parameters,
- storage of plant protection products, and
- disposal of empty containers.

This area showed some favourable changes in Mexico, but to a lesser extent than in the personal safety area. The changes that persisted after the intervention programme ended include not repacking the pesticide package, storing pesticides out of children's reach, and improved disposal of empty containers.

The safe storage of crop protection products was at a high level in both regions in India at the baseline, and that continued through 1997. The safe disposal of pesticide containers showed no improvement in either area. The use of a measuring glass was at a high level in 1992, fell in 1996, and then rose again in 1997 back to the level at the baseline. The intervention programme ensured that a good practice more or less continued.

The baseline study in Zimbabwe revealed a high level of knowledge in the area of storage of chemicals and disposal of empty containers. Although there was no change in the first regard, the correct disposal of empty pesticide containers improved in the test area. A cotton pest calculator was introduced in an attempt to improve the use of the optimum quantity of pesticide. The result was not noticeable, however, as this tool needs a more comprehensive introductory programme, including field training working with heavy pest infestation. Such training was not possible during the project period.

Hypothesis: Through communication and training we will improve practice in:

- identification of pests and beneficial insects,
- selection of suitable product,
- determination of correct dosage,
- usage of suitable equipment,
- correct timing of application, and
- proper application techniques.

There were relatively few detectable effects in Mexico in the area of pest identification and product selection. Knowledge of beneficial insects improved, and focus group sessions detected more rational attitudes towards pesticides. However, no yield or productivity effects were detected by the KAP studies even though demonstration plots managed by project technicians produced superior crop yields.

In India, inspection of plants before taking a decision to spray and the decision regarding the pesticide to be used both improved significantly as a result of the intervention, but this change did not continue. Knowledge of beneficial insects amongst farmers and spraymen rose substantially but it dropped later (though it remained at a higher level than at the baseline). The practice of determining the correct dosage improved substantially and more or less continued in the project area. The commencement of spraying operations before 9:00 a.m. (when it is cooler and therefore there is less evaporation, making use more effective) rose significantly among farmers and spraymen in the project region, but this positive change was not sustained.

It is not easy to summarize the improvement in this area due to the project in Zimbabwe. Farmers received chemicals in the past through a credit scheme. The impact of a liberalized market on the use of suitable products remains to be seen. Knowledge of beneficial insects remained at a moderate level. But only half the farmers reported that they keep up with regular maintenance of equipment.

Hypothesis: Improvement of farmers' economics will facilitate their adoption of messages on safety.

It is difficult to prove this hypothesis with a KAP or observation survey. Evaluation should instead be based on the results of qualitative surveys and the experience of local project staff. Still, it seemed that in Mexico the demonstration of more-productive crop protection techniques on the plots managed by the project staff sparked farmers' interest and raised their level of trust, which in turn facilitated communication. More rational attitudes towards pesticides affected farmers' perception of both effectiveness and safety issues. Otherwise, however, there does not seem to be a strong link between these issues, as safer practices were adopted despite a lack of noticeable improvements in effectiveness.

Qualitative research studies in India indicated that there was widespread appreciation among farmers that Novartis had put considerable effort into improving farmers' health by teaching them how to enhance their safety as well as improve their crop yields through the effective use of pesticides. This credibility of Novartis could be considered as contributing in a large measure to the positive changes noted in the adoption of safety measures and the consequent improvement in the health of the farmers and spraymen in the project area, as reported during informal interviews during a visit to the region in 1998.

The continuous changes in the climate in Zimbabwe from year to year and a severe drought in 1995/96, which meant the loss of food as well as cash crops, had a considerable negative impact on farmers' economic situations. In general, available funds are used first for basic needs (daily operating expenses, for example, or schooling for children) before investments in safer use of pesticides are made. It is too early to tell if the liberalized cotton market will improve the financial situation of cotton farmers there enough to allow them to purchase adequate safety gear and spraying equipment.

Overall, therefore, the interventions did have a positive impact. We learned some important lessons during the course of the project. The most fundamental one was that messages need to focus on practical, basic, ready-to-use, but effective recommendations. Suggesting the use of impractical or expensive items or habits can dilute the overall message about safety precautions. Indeed, a highly technical and expensive approach is not needed to improve safety. The good news is that a small number of simple changes make a big difference and this, according to our experience, is feasible. In addition, the mix of communications media used in each country kept being refined during the project, reinforcing the point that social marketing campaigns need to be tailored to specific locations.

Special efforts were made to include in the communications campaign children of the farmers who use pesticides, which was also innovative. In Mexico, a playbook and cartoons were used, while in Zimbabwe children were encouraged to attend the plays put on by the project. A series of programmes for rural schoolchildren was introduced in India. This not only helped spread the message to the parents, it also served to prepare tomorrow's farmers to use pesticides safely and effectively. The project's use of these communications media could be called ground-breaking.

Despite the increase in the number of farmers adopting improved practices, a large number still did not do so even though they were aware of the health risks. There are at least three possible explanations for the persistence of the phenomenon. The first is that the "early adopters" or those most susceptible to change may have adopted some of the changes, leaving the laggards yet to be convinced. (This will necessarily take more time.) The second is that the producers in the regions are very poor so they are risk-averse and cannot take chances about the possible consequences of change, especially if the changes involve financial outlays. For example, the purchase of protective gear fell sharply after a subsidy from the project was removed in India; conversely, though, few farmers in one area used gloves (a recommended safety precaution) even though they were made available at no cost by a major pesticide company.

A third reason for farmers not taking precautions even though they were aware of the importance of doing so was that "external forces" overwhelmed rational behaviour. Thus, the drought in Zimbabwe and a sharp devaluation in Mexico disrupted all farming activities, including those related to the use of pesticides. These may well have been short-term disruptions, however, so their long-term effects are still open to question.

One somewhat sobering result of the project is that social marketing campaigns have to be done on a sustained basis, as farmers tend to fall back into undesirable attitudes and habits after some time. Change cannot be maintained with time-restricted interventions only. There is a pronounced need for ongoing intervention to ensure persistent change.

Another important lesson is that safety messages in isolation are not likely to have a big impact, but when delivered in conjunction with an effectiveness message, the safety point can be much more appealing to the farmer. (The farmer, however, analyses and assesses the package, and discards components according to a number of factors). Messages that do not exclusively focus on the farmers but also involve their social environment (the family, peers, and so on) are more successful, as they trigger additional pressure and heightened awareness about the issue. An attractive intervention programme that seems to be in the overall interest of the farmer will create a ripple effect over and above the intervention group or area, leading to a broader impact on society.

In conclusion, the project draws attention to the fact that if farmers were to take a series of relatively simple steps, they could reduce their exposure to pesticide-related health risks. At present, many if not most farmers give low priority to "safety", and many have not adopted the necessary precautions to reduce health risks. Some procedures may well be made more acceptable to low-income farmers—for example, by developing and subsidizing the sale of both cheap and comfortable clothing that can provide adequate dermal protection. In the main, though, it appears that there are few if any easy ways to promote change among large numbers of poor smallholders.

There will have to be a continued reliance on sustained efforts such as some of those incorporated in this project. But all available experience indicates that there are limits to the extent to which changes will be adopted within a generation. Even the best and most sustained efforts run into the paradoxical situation that not everyone who can adopt relatively simple modifications in behaviour will actually do so, even when it is shown that the changes are in the person's long-term best interests. Given that, any pesticide manufacturer that cannot guarantee the safe handling and use of its toxicity class 1A and 1B products should withdraw those products from the market. At the same time, since in all likelihood pesticides will continue to be an essential crop protection tool in the years ahead, there is a continued need to get farmers to adopt the most important risk-reducing procedures.

1

Background and Overview

The ability of the world's farmers to produce food has improved dramatically over the past four decades. Per-hectare yields of maize, rice, and wheat nearly doubled between 1960 and 1994, for instance, and similar progress was made in other crops. These achievements were largely due to technological progress in the form of improved varieties, irrigation, fertilizers, and a range of technologies that aided farmers' management of crops and resources.

The breadth of this technological progress becomes obvious if we look at what has been achieved in India over the past 25 years. In 1991–93, India produced on average 196 million tonnes of grain a year, with an average yield of 1.98 tonnes per hectare, but in 1961–63, the yield figure stood at 0.95 tonnes per hectare. If India were still using the technology of the 1960s in the 1990s, the nation would need 208 million hectares of arable land—116 million more hectares than were available in 1961–63. Thus if the yield per hectare had not doubled, achieving the results recorded in 1991–93 would have required doubling the land under cultivation—a sheer impossibility without causing an ecological disaster by converting the last remaining forests to cropland.

The technological package that allowed this progress—the Green Revolution—had these key components:

- modern seed varieties that had, among other traits, short maturation times, insensitivity to variations in daylight, and the ability to convert the application of fertilizer into high crop yields rather than stem and leaf growth;
- fertilizer;
- irrigation;
- mechanization (partly); and
- insect and weed control.

This chapter is based on material prepared by Jost Frei, Klaus M. Leisinger, Peter Rieder, Karin M. Schmitt, and Hema Viswanathan.

Some scientists and policymakers see substantial negative consequences to the Green Revolution, while others see huge benefits. Overall, the best judgement today is that the Green Revolution had a positive socioeconomic effect. While landowning households often benefit most from the direct income effects of agricultural growth, the landless as well as smaller, food-deficit farmers often benefit most from the indirect effects, such as the generation of off-farm employment. Even where the Green Revolution made the rich richer (because they could use the new technologies earlier and on better land with better inputs and less expensive credits), the poor also benefited—over time becoming less poor as agricultural modernization proceeded.

Where small farmers were supported by agricultural extension services, where they had access to land, inputs, and credit—in other words, where there was "good governance" in the sense of an agricultural development framework designed to assist the endeavours of the small farmers—they were able to benefit a great deal and early. Rapid productivity gains have in general decreased food costs and improved food security, particularly for vulnerable sections of society. The urban poor have been important beneficiaries of this.

Of course, now scientists are talking about the next revolution—one based on biotechnology. Transgenic plants may be introduced in the years ahead. But there are still many open-ended questions about whether these new technologies will be a major factor in increasing agricultural output in the developing countries. Consequently, strategies for increasing yields will continue to rely, in good measure, on refinements of technologies now in use. Unlike the preceding 50 years, though, when the main strategy for development was to increase inputs into agriculture and to raise the capacity for production, the emphasis will have to change. The economic environment has changed. The competition for capital, moves towards globalization, privatization, opening up of markets, and freer trade all mean that there can be less reliance on adding inputs to raise output and to expand capacity; instead, greater attention will have to be paid to using available resources—including pesticides—more efficiently than in the past.

This is not the place to refight the battle of the Green Revolution, which has been well waged elsewhere. But it is important to acknowledge that pesticides (mainly herbicides, fungicides, and insecticides) have been a major source of criticism about the Green Revolution—and a major campaign issue against multinational corporations and their activities in developing countries. As crop yields were increasing due to better varieties, fertilizer, and irrigation, farmers shifted away from multiple cropping and crop rotations in order to grow high-yielding varieties in monocultures and continuous cultivation. This made crops more attractive to pests, diseases, and weeds, and hence increased the need for crop protection.

This criticism of pesticides and their use in developing countries is an important part of the background for this study.

Pesticides: Risks and Benefits

Crop management can, should, and often does involve biological, mechanical, and cultural methods, and so is a much wider concept than just chemical crop management. Because many governments have been promoting standard packages with seeds, fertilizer, and pesticides, however, and because there have been government subsidies and substantial marketing efforts by the chemical industry, and due to its effectiveness, there has been a marked bias towards chemical crop protection. Although nonchemical approaches to crop management may have some advantages over chemical ones, and although fine-tuning of integrated pest management may make an additional wide range of effective methods available, the discussion here is restricted to chemical crop protection products and their application and use, which was the focus of the research in this project.

The discussion of the use of such products in developing countries falls into two adversarial camps. First, those who see more risks than benefits point to hazards to human health, wildlife, and the environment, as well as to high costs and low efficiency due to pests' ability to build up resistance and the destruction of their natural enemies. Given the prevailing conditions in developing countries (illiteracy, ignorance of side effects, and so on), the safe and proper use of chemical plant protection products cannot be guaranteed. Gordon Conway summarizes the criticism thus:

> Pesticides not only cause or aggravate pest problems..., they contaminate the environment and may have serious consequences for human health. Their effects on wildlife are well documented. Organochlorine insecticides, such as DDT and dieldrin, which were in common use in the 1950s and 1960s, are highly persistent. In the developed countries they caused dramatic declines in birds of prey, such as the peregrine falcon, and had less visible, yet equally serious, consequences on a great variety of other wildlife. Similar effects must have occurred in the developing countries, although the evidence is not as complete. (Conway, 1997, p. 87)

There is no doubt that pesticides can cause and have caused harm to nontarget organisms. And there is little doubt that the inappropriate use of these products can be downright counterproductive. Yet the extent of the harm may be in the eye of the beholder. Both opponents and defenders of these products wield their statistics, enriched by anecdotal evidence. The opponents seem to have been either the more audible or the more credible party to talk about the issue. Public interest groups, in particular, have gained considerable media coverage and influenced the information databases of parliamentary investigating committees and social science institutions. Today, broad sectors of the public hold a generally sceptical or even adversarial attitude towards chemical crop protection and its manufacturers.

On the other hand, those who see more benefits than risks in the use of pesticides point to substantial gains due to the prevention of crop losses. There is a surprising paucity of reliable empirical data on pest-induced losses under realistic field conditions in developing countries. Even specialists at international agricultural research centres come up with estimates for specific crops and specific countries that vary tremendously, sometimes by a factor of 10. One systematic analysis found that pathogens, animal pests, and weeds can cut the potential production of practically all food crops nearly in half (Oerke *et al.*, 1994).

Even if we accept the notion that it is not appropriate to compare a "worst case" with a "best case" scenario, the fact that pest-induced production losses may top 50% is significant. Where crop losses before and after harvest are prevented, more of the potential harvest becomes actual harvest and thus available to farmers. As the available inputs are better used, less land is required to produce a given quantity of food or fibre, so crop protection products are a land-saving technology. Of course, this does not mean that other ways to conserve land are not also appropriate.

Factors Clouding the Judgement on Pesticides

Numerous issues are making it difficult to form a sound judgement on pesticides. One of the most important is the fact that, according to the *Pesticide Manual*, just under 800 active ingredients of pesticides are now in use, each with different characteristics. A pesticide can be an old-fashioned, extremely dangerous, nonselective product and, from the point of view of today's professional, not a state-of-the-art item. Many of these products—of which some kilos per hectare must be applied—have been banned for years by regulatory authorities in industrial countries, yet they continue to be produced and sold at very low prices in many developing countries, most often by national companies. In contrast, a pesticide can be a very modern product—highly selective and not toxic to humans or other living organisms that are not the target of the chemical. The best of these products can be applied in small quantities (a few grams per hectare) and are highly effective. But they are most often considerably more expensive than the old products still around.

Another issue that is making a sound judgement hard to come by is that national health statistics do not clearly differentiate between occupational health problems and intentional wrong use (suicide or murder). This differentiation is essential, however. Considerable personal experience and anecdotal evidence indicates that there is widespread misuse of pesticides for suicide and crimes.

During a Steering Committee visit to southern India, for example, doctors at the emergency station of a hospital reported that some 400 cases of poisoning from pesticides had occurred during the past five years. Apart from three attempted murders, they all turned out to be attempted suicides with organophosphates, divided equally between women and men. Of the total, 20% ended fatally; none of them happened in connection with a person's occupation. Other physicians in the region likewise stated that cases of pesticide poisoning are primarily suicide attempts and only rarely due to occupational accidents.

The attractiveness of pesticides for this is obvious. To be an effective means of suicide or homicide, a poison must be efficacious and quick-acting, cheaply and fairly readily available, not arouse suspicion when purchased, and not cause retching when swallowed. Each income class, it was reported, resorted to different means of suicide. The poorest used cardiotoxic oleander leaves; those in the middle-income class used crop protection products; and the rich, sleeping tablets.

If pesticides are indeed used for deliberate actions such as suicide or murder, then the thesis that the hazard of agricultural chemicals is mostly unknown to poor and illiterate people in poor nations is not credible. Although physical harm or even death due to deliberate pesticide intake is undoubtedly a tragedy, it would be inappropriate to mix this up with occupational health issues. The ongoing efforts of the World Health Organization to shed more light on this issue are therefore highly laudable.

Few technological issues have caused as much debate as the use of pesticides. The assessment of the contribution that these crop protection products can make towards producing affordable food in developing countries, the evaluation of the damage they have caused to humans, to natural enemies, and to the wider environment, and the balancing of the two components is not "simply" an academic task—an exercise where facts and figures are collected and rationally assessed. Evaluations of this kind are subject to a great variety of interests and value judgements by a multitude of stakeholders. On the basis of the information available plus selected evidence chosen by one of the stakeholders, some authors consider pesticides to be an economically promising technology, while others perceive them as a threat to development in poor countries.

As in many cases of public policy, the theory of constructivism holds true: there is no such thing as *the* reality. Instead, as the result of differing value judgements, world views, and personal experiences, there are different subjectively perceived realities. Individual observers regard what they are able to see or would like to see from their viewpoints as uniquely real, and they assess their perceptions according to preconceived ideas and basic assumptions.

In the case of pesticides, differing realities and hence pluralism of opinion are more complicated, as most people confronted with the issue are not specialists in biology, toxicology, or ecology and hence have to believe what others say or the media discuss. Wild stories about the poisoning of innocent children or about marketing managers who lack morals and professional responsibility because they oversell in an environment of poverty are more likely to be taken up by media than reports about slow but steady progress with regard to responsible pest control and product stewardship efforts. As most people in industrial countries have easy access to food and little knowledge of what pests can do to crops, they perceive pesticides as having high risk without being able to assess the benefits.

As we live in a world with very heterogeneous social systems, with a multitude of value judgements and interests, we must live with these deviating evaluations. On the one hand, there are obvious agricultural benefits from the rational use of crop protection products. They have a significant potential to increase production and productivity and hence contribute towards food security and a better quality of life. On the other hand, there are a number of economic, social, and ecological risks and costs. Crop management with chemical agents is not without problems, particularly when they are used indiscriminately. The *International Code of Conduct on the Distribution and Use of Pesticides* adopted by the Food and Agricultural Organization of the United Nations includes recommendations in considerable detail about all important aspects of an appropriate duty of care. The code is supported by the Global Crop Protection Federation, which all globally active research-based companies belong to; companies must adhere to the code in order to join the organization.

In this respect, pesticides are no different from other forms of technological progress, which, as the German theologian Helmut Gollwitzer once said, is "nothing other than the unremitting struggle to secure its positive aspects, learning to live with the dangers that come with it and surmounting the impairments it causes" (Gollwitzer, 1985, p. 142). Exactly what constitutes the "positive aspects", "dangers", and "impairments" in a given case is the stuff of dispute—as is the definition of an "acceptable risk". The attractiveness and desirability of a certain effect of technological progress is very much a function of individual value judgements.

The result of this evaluation can be utterly irrational: perceived risks can be much lower or much higher than actual risks. Although most people in industrial countries are willing to accept a technology—the automobile—that is contributing to global warming, that kills about 50,000 persons a year and maims another half-million in the United States alone, and that adds nothing vital to their lives except the added convenience of personalized fast travel, the use of pesticides is perceived by many as too risky to be acceptable.

The important question cannot be, Pesticides: yes or no? Any serious analysis must ask a number of differentiating questions such as, What products? Under what social, educational, and ecological circumstances? For what purposes? With what precautions? Are there cost-effective alternatives readily available? Efforts to promote the safe and effective use of pesticides can make a big difference—at least, that was the hypothesis of this project.

Social Marketing and Studies of Knowledge, Attitudes, and Practices

Sharpening society's awareness of a problem such as misuse of pesticides is necessary, but it is by no means sufficient for bringing about changes in attitudes and behaviour, as these are shaped by habits, interests, feelings, and beliefs, among other factors. That is why social campaigns conceived simply to educate or admonish often turn out to be relatively ineffective.

These limitations, along with the success of advertising techniques in the commercial world, provided the impetus for the development of social marketing. Philip Kotler and Gerald Zaltman used this term in 1971 to describe the use of marketing principles and techniques to advance a social cause, idea, or behaviour (Kotler and Zaltman, 1971). Since then, the term has evolved to encompass social change management programmes that aim to make a social idea or practice more acceptable to one or more targeted audiences (Kotler and Roberto, 1989).

To this end, social marketing makes use of methods from the commercial sector: setting measurable objectives, doing market research, developing products and services that correspond to genuine needs, creating demand for them through advertising, and finally marketing through a network of outlets at prices that make it possible to achieve the sales objectives.

Social marketing aims to "reach" one or more target groups in order to initiate and bring about changes in their ideas and behaviour. The starting point, therefore, is getting to know the target audience thoroughly through market research: the audience's social and demographic makeup (economic status, education, age structure, and so on), its psychosocial features (attitudes, motivations, values, and behavioural patterns), and its needs.

Any campaign must take these factors into consideration and also be compatible with the cultural and religious traditions of the target group or groups. Market research is crucial not only in the planning phase of the programme but also during its implementation, as social marketers have to be aware of and responsive to the target groups' changing needs.

The project was started in the nature and spirit of an experiment in social marketing—in this case, the idea being marketed was the safe and effective use of pesticides. Adoption of practices that were both safe and effective would require that certain current practices be abandoned in favour of new ones. This in turn would require that the target audience recognize and accept the poor effectiveness or hazard of their current practices and be willing to make the effort required to change.

For changes such as these to occur, people must see a personal benefit from the new behaviour. The persuader (or social marketer) must understand the circumstances and constraints under which target audiences operate, their priorities and needs, and their knowledge level and beliefs. Persuaders must speak with empathy and understanding and in the language, idiom, and terminology of the audience if they are to have any hope of holding attention, persuading, and converting.

Practices can at best be as good as knowledge: while people can knowingly and deliberately behave in a way that is different from or less than their knowledge, they cannot do better than the best they know. A person's knowledge therefore defines the upper limit of his or her practice. Further, practices are influenced by personal and business objectives, priorities, needs, and beliefs, which in turn are reflected in attitudes. Thus, both knowledge and attitudes have a role to play in determining practice.

The first step in the project was therefore to understand small farmers in developing countries; to recognize their needs, aspirations, and priorities; to observe their practices; and to look for patterns of behaviour in terms of their use of plant protection products. A study of knowledge, attitudes, and practices (KAP) designed to explore, understand, measure, and profile attitudes and behaviour thus was a logical starting point for the project.

The main objectives of the KAP study were to:

- acquire insight into the knowledge, attitudes, and practices of the target groups as they relate to crop protection and the use of pesticides;
- gauge the significance of knowledge and attitudes for the rational and effective use of crop protection products;
- establish indicators that can be used to measure future changes in KAP; and
- analyse differences and similarities in three geographically and socioculturally dissimilar study sites, with a view to determining whether standardized modes of intervention for increasing the safe and effective use of pesticides are possible.

The step-by-step process of evolving a social marketing strategy has been described as:

- defining the problem and setting objectives,
- identifying the target audience,
- defining the proposed behaviour change,
- identifying the resistance points,
- assessing media availability,
- designing the product, and
- choosing the distribution systems.

The project planners had defined the problem—as, indeed, had critics of chemical protection of plants—as farmers in developing countries using pesticides in a way that is both unsafe and less than economical. But this was just one side of the problem definition and therefore only one half of the story. The other half was the definition of the problem from the point of view of small farmers. It would have been wrong to assume that they shared the same concerns as the agrochemical industry or its critics. Our baseline KAP studies in fact showed that small farmers had a whole host of other concerns, and that danger from pesticides was low on the list of their immediate worries. To cite but one example, in some societies it goes against the grain for men to put on protective clothing when applying pesticides, as the attitude is that "only weaklings wear such outfits."

Underlying Questions and Hypotheses of the Project

Given this background of differing views on pesticide use, the project was undertaken to clarify the following questions:

- What factors hinder the safe and effective use of crop protection products in developing countries?
- What sort of steps must or can manufacturers take, in collaboration with other institutions and organizations (such as agricultural extension services), to eliminate these constraints on safety and effectiveness?
- In a given sociocultural context, what communication methods are suited to furthering rational and safe use of crop protection products?

In the light of the information on pesticides described at the beginning of this chapter and the definition of the problem by the Steering Committee, we established the following basic facts on safety and effectiveness, along with the hypotheses to be tested in the Risk Fund Project.

> Fact: Reduction in dermal contamination leads to reduction of possible health problems.

Hypothesis:
Through communication and training we will improve practice in:

- skin protection,
- preparation of spray solution,
- washing of body and work clothes,
- spraying and application, and
- maintenance of spraying equipment.

> Fact: Proper handling of plant protection products limits the negative impact on the environment.

Hypothesis:
Through communication and training we will improve practice in:

- optimization of quantity of plant protection products used and of spray parameters,
- storage of plant protection products, and
- disposal of empty containers.

Fact: Effective handling and use of plant protection products improves the economic situation of the farmer.

Hypothesis:
Through communication and training we will improve practice in:

- identification of pests and beneficial insects,
- selection of suitable product,
- determination of correct dosage,
- usage of suitable equipment,
- correct timing of application, and
- proper application techniques.

Fact: The farmer values economics more than safety.

Hypothesis:
Improvement of farmers' economics will facilitate their adoption of messages on safety.

Basic Design of the Project

The selection of countries to be included in this project was guided by a desire to carry out the social marketing experiment in three nations that represented similar economic development but widely disparate sociocultural environments as well as agricultural practices. India, Mexico, and Zimbabwe were chosen to participate in the project. Combined, they are fairly representative of agriculture in the developing world. Novartis affiliates in each country helped set up local project organizations that took responsibility for the baseline KAP study and for implementation of the communication campaign that followed it. The Indian Market Research Bureau was responsible for development and evaluation of the KAP studies.

Certain common characteristics were found to a greater or lesser extent in all three countries that met the criteria laid down for country selection:

- a large population of small farmers with low levels of formal schooling;
- inadequate knowledge of safe and effective use of crop protection chemicals;
- unsafe practices adopted in the usage of pesticides, which exposed users to avoidable hazards as well as causing environmental pollution;
- an increasing consumption of crop protection chemicals in order to enhance agricultural production and productivity;
- climatic conditions that are not conducive to wearing protective clothing and gear as recommended;
- government extension services that have declined under outside pressures; and
- inadequate or nonexistent enforcement of legislation to control usage of pesticides.

Specifically, the projects were located in the following areas (see Figure 1.1):

- in India, in the Coimbatore district of the southern state of Tamil Nadu,
- in Mexico, in Villaflores and Cintalapa in the southern state of Chiapas, and
- in Zimbabwe, in Sanyati and Nemangwe in the centre of the country.

Further details of these different areas are provided in Chapters 3, 4, and 5. In each case, a control area was designated in which only the surveys were undertaken. In contrast, in the areas designated for intervention, the baseline KAP study was followed by a three-year communications and training campaign tailored to the particular area. Project results throughout the book for the baseline KAP study, observation studies, and an endline KAP survey are thus reported in terms of both control and intervention areas.

The target audience for the intervention component was identified as the small-scale cotton-growing farmer (in India and Zimbabwe) and the small maize-growing farmer (in Mexico). Empirical evidence had shown that these farmers were regular users of pesticides but that their illiteracy and poverty made it very unlikely that their methods of use were well planned or followed application instructions or user safety precautions. The heavy use of pesticides per hectare also suggested possible indiscriminate use. Anecdotal evidence had repeatedly suggested that usage practices were often careless, if not downright hazardous.

In India, there was an additional target category defined as the sprayman. This is someone who hires out labour on a contractual basis to farmers who might want to have their fields sprayed. As a result, during peak season, spraymen could spend several hours a day, several days in a row, spraying pesticides. Their level of exposure to risk was therefore high and they constituted an important audience.

As a rule, the communication channels in any social marketing campaign selected should be ones the target audience comes into contact with on a regular

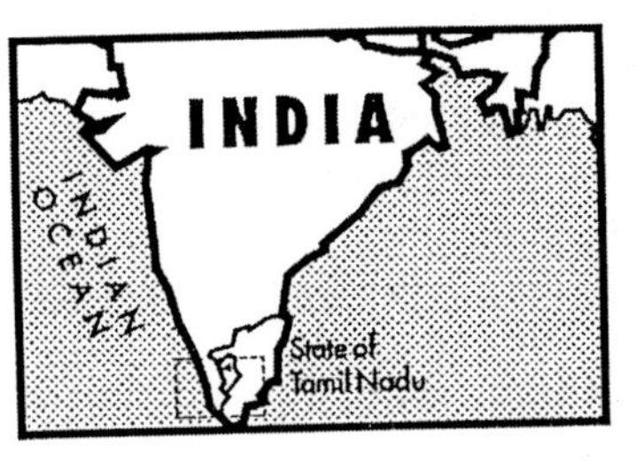

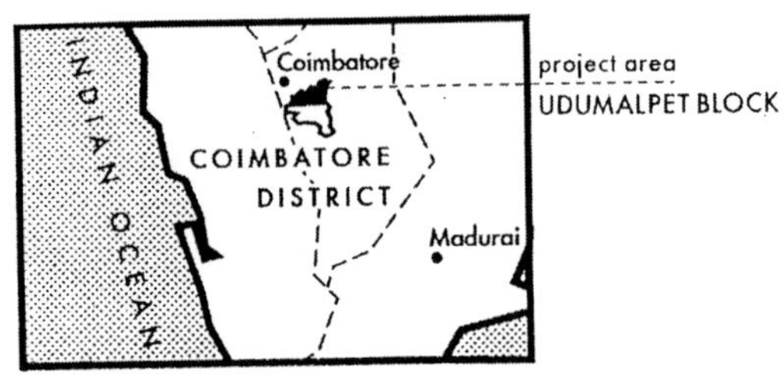

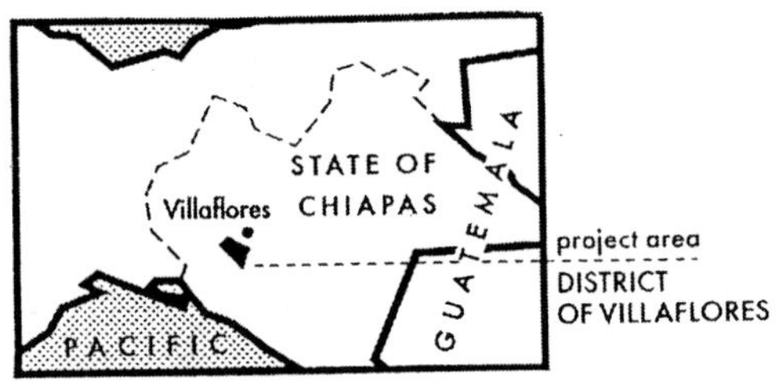

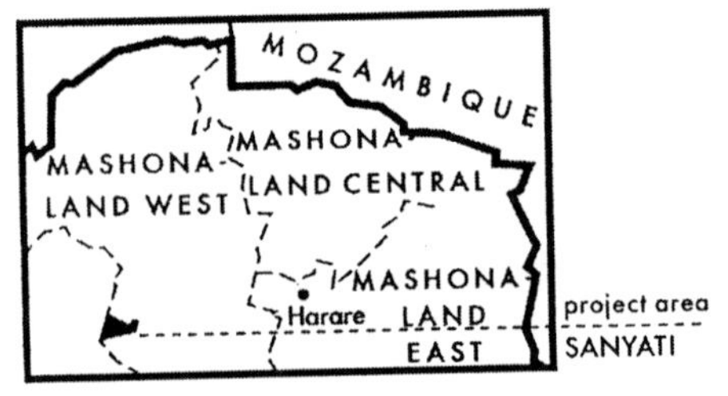

Figure 1.1 Locations of the Novartis project

basis as well as perceives as being credible, since familiarity with a medium and with the performers makes it easier to get the message accepted. Projects that use media with entertainment value (films, soap operas, radio plays, music, theatre, comics, and so on) are particularly successful. Members of the target group can identify with the heroine or hero or with a well-known idol, and this has a motivating effect in the desired direction of change.

Before our communication campaigns were launched, it was therefore important to consider the availability of media for reaching small farmers as well as to identify the media that were most attractive and credible. In Mexico, radio has both audience appeal and reach, and was used to reinforce messages delivered during special farmers' meetings. Fortunately, the option of localized radio broadcasts made it possible to isolate communication to the intervention area without having the messages reach the control area.

In India, in contrast, films are an important source of entertainment for the overburdened masses, who seek it as a form of escape from their daily drudgery. Films have a hold on most people that is difficult to explain. Screening a film in a village is therefore a sure way to draw crowds and to provide messages along with entertainment. The use of television and VCRs allowed the film to reach all test villages without spilling over into neighbouring regions.

In Zimbabwe, folk theatre and the oral rendition of stories and messages is a lively, vibrant, and yet ancient tradition. It has instinctive and almost reflex appeal, and this alone made the medium the right choice. In addition, it was versatile, flexible, and interactive, and could reach out to the most remote corners of a given region.

Following baseline KAP studies in 1992 and 1993, the communication campaigns were designed and launched in 1993/94. The way we measured their impact is described in Chapter 2. We found that performing a "real world" experiment posed some unique methodological challenges.

Bibliography

Angehrn, B. (1996) *Plant Protection Agents in Developing Country Agriculture: Empirical Evidence and Methodological Aspects of Productivity and User Safety.* Thesis No. 11938, Swiss Federal Institute of Technology, Zurich.

Barker, R., Herdt, R.W. and Rose, B. (1985) *The Rice Economy of Asia.* Resources for the Future, Washington, DC.

Barrons, K.C. (1981) *Are Pesticides Really Necessary?* Regnery Gateway, Chicago.

British Medical Association (1987) *BMA Guide: Living with Risk.* John Wiley, Chichester and New York.

Bull, D. (1982) *A Growing Problem: Pesticides and the Third World Poor*. Oxfam, Oxford.

Conway, G. (1997) *The Doubly Green Revolution. Food for All in the 21st Century*. Penguin Books, Harmondsworth, Sussex, UK.

Gollwitzer, H. (1985) *Krummes Holz - Aufrechter Gang: Zur Frage nach dem Sinn des Lebens*. Auflage, Christian Kaier Verlag, Munich.

Gunn, D.L. and Stevens, J.G.R. (eds). (1976) *Pesticides and Human Welfare*. Oxford University Press, Oxford.

Hayes, W.R. (1982) *Pesticides Studies in Man*. Williams and Wilkins, Baltimore.

Knirsch, J. (1996) Pesticides and global food security: an examination of chemical industry arguments in support of "feeding the world through chemical plant protection". In: PAN International. *Growing Food Security: Challenging the Link between Pesticides and Access to Food*. London, pp. 6–11.

Kotler, P. and Zaltman, G. (1971) Social marketing: an approach to planned social change. *Journal of Marketing* 35 (July): 3–12.

Kotler, P. and Roberto, E.L. (1989) *Social Marketing: Strategies for Changing Public Behaviour*. Collier Macmillan, London.

Leisinger, K.M. (1986) Multinational companies and agricultural development: a case study of "Taona Zina" in Madagascar. *Food Policy* 12: 227–241.

Leisinger, K.M. (1987) Towards a green evolution. *Development Forum* 15 (2): 8–9.

Levine, R.S. (1986) Assessment of mortality and morbidity due to unintentional pesticide poisoning. Paper presented to the Consultation on Planning Strategy for the Prevention of Pesticide Poisoning. World Health Organization, Geneva.

Loevinsohn, M.E. (1987) Insecticide use and increased mortality in rural central Luzon. *Lancet*, June 13: 1359–1362.

Maturana, H.R. (1985) *Erkennen: Die Organisation und Verkörperung von Wirklichkeit*. Viehweg, 2nd printing, Braunschweig.

Oerke, E.-C., Dehne, H.-W., Schoenbeck, F. and Weber, A. (1994) *Crop Production and Crop Protection: Estimated Losses in Major Food and Cash Crops*. Elsevier, Amsterdam.

Rola, A. and Pingali, P. (1993) *Pesticides, Rice Productivity, and Farmer's Health: an Economic Assessment*. International Rice Research Institute, Los Baños, Philippines.

Turnbull, G.J. (ed). (1985) *Occupational Hazards of Pesticides Use*. Taylor and Francis, London.

U.N. Food and Agriculture Organization (1986) *The International Code of Conduct on the Distribution and Use of Pesticides*. Rome.

U.N. Food and Agriculture Organization (1995) *Production Yearbook 1994*. Rome.

Utsch, H. (1991) Pesticide poisoning in a rice and vegetable growing area of Sri Lanka. Inaugural dissertation. University of Zurich.

Watzlawick, P. (1989) *Wie wirklich ist die Wirklichkeit?* Piper, München.

World Health Organization (1990) *Health Impact of Pesticides Used in Agriculture*. Geneva.

Yudelman, M., Ratta, A. and Nygaard, D. (1998) *Issues in Pest Management and Food Production: Looking to the Future*. Food, Agriculture, and the Environment Discussion Paper No. 25. International Food Policy Research Institute, Washington, DC.

2

Conducting Research in a Changing World

The research done in the context of this project had several goals:

- collecting firsthand data on the actual situation regarding pesticide use, including knowledge, attitudes, and practices (KAP), the state of pesticide-related illness, and economic indicators;
- testing the hypothesis that the situation could be improved through a communication, education, and training programme; and
- evaluating how the improvements (if any) were achieved.

The methodology used in this research is discussed in this chapter as we address the question, How can we measure the effects of the programme on KAP and, eventually, on farmers' health?

The answer to this question involves a number of considerations, ranging from the definition of adoption, to the influence of attitudes on the reporting of illness, to the mathematical formula used to calculate differences between adoption levels. Many of these issues arise because any effort to measure the impacts of a social marketing campaign is a moving target. These impacts occur over a period of several years during which many other changes take place in the real world where the "experiment" was conducted. The methodology used was designed to deal with external influences, but the speed and magnitude of those changes was a surprise. As the intervention period was extended for various reasons (see Chapters 3, 4, and 5), external factors became even more important. As a result, adoption of promoted behaviour is sometimes small compared with other effects, and sometimes follows unexpected patterns. It therefore is the overall picture—the compound effect on a number of variables—that should dominate the analysis.

This chapter is based on material prepared by Beda Angehrn, with contributions by Al Imfeld.

Bearing this in mind, the more technical methodological aspects are outlined in this chapter.

Measurement is difficult in the first place because of problems in determining when a new behaviour has been "adopted". Individual adoption of some new behaviour is the degree to which an innovation is implemented by an individual, while aggregate adoption refers to the degree to which a population group implements the innovation. Both these definitions implicitly assume that adoption involves the choice of either adopting or not. This simplified view ignores the fact that an individual may adopt only certain elements of an innovation or may adjust parts of it to meet current needs. Similarly, an innovation may involve a "package" of several elements that individuals may adopt jointly or on their own. In addition, individuals within a population may make different uses of an innovation to different extents—a dimension that is not captured in such a measure as the proportion of individuals that adopted the innovation.

Measurement is also difficult because of different unspoken attitudes among researchers concerning the questions included in the survey. In the case of this project, for example, some assumed that sprayers would be "macho" personalities, that farmers would be concerned only with effectiveness, that farmers would blindly follow written or oral instructions on the use of pesticides, or that farmers reveal their attitudes through the actions they take automatically. These hidden attitudinal differences among our own team of researchers about the questions asked during the project surveys clouded the interpretation of the results more than we expected.

The process of adopting new behaviour is triggered by the availability of information about a new technology. The process may start slowly, gain speed, and then tend to "flatten out" with time as individuals have full information about the technology and its potential. The level that adoption would achieve in the long run is called "final adoption". In reality, adoption processes overlap as technologies become continuously available. Adoption of the newest technology may reduce or accelerate adoption of earlier ones. Indeed, it might be best to talk of adoption cycles rather than adoption of a single technology. Changes in the conditions within which adoption takes place (prices, for instance) also affect adoption. And the effectiveness with which a new technology is used may vary with time as a result of "learning by doing".

The adoption process is the result of a number of decisions made by individuals at different points in time, and of the relations between them. To understand adoption, we must understand both the decisionmaking problem and its changes over time, which are the result of a learning process that incorporates prior perceptions and knowledge along with recent information on the performance of the technology or technique. The variables that determine the decision may be categorized as knowledge (information) and attitudes (values). The decision produces a practice that in turn will influence knowledge and attitudes through the experiences it creates.

In the context of agriculture, a large body of literature has examined the factors that favoured or hindered the adoption of "modern" production techniques or elements of the Green Revolution, such as high-yielding varieties, chemical fertilizers, and pesticides (see, for example, Feder, Just, and Zilberman, 1985). Such factors include the quality of the available information, the decisionmaker's skills in processing this information and thereby reducing subjective uncertainty about the benefits and costs of the innovation, the profitability of the technology and its variability, the relation of this variability to the farmer's capacity and willingness to bear risks, access to credit, and tenurial arrangements.

Beyond the problem of defining "adoption", it is important to consider the possible effect of the research methodology itself. As noted in Chapter 1, we sought to examine and quantify the effect of communication, education, and training programmes on the adoption of safer and more effective pesticide use practices. This posed four basic problems:

- identifying the effect of the programme,
- choosing suitable variables to represent the goals of the programme,
- choosing a suitable model for estimating the net effect of the programme, and
- choosing a suitable level of aggregation of the analysis.

We discuss these four issues in the remainder of this chapter.

Identification of Programme Effect

To determine the effect of any programme, it is necessary to separate the effects of other unrelated factors, such as the influences on adoption of any innovative technology. We noted three groups of such external factors:

- economic factors—that is, factors influencing the profitability of pesticides. These include such external changes as shifts in crop or pesticide prices due to market forces or regulation, climatic factors affecting yields, and the prevalence or pressure of a pest.
- information factors—that is, changes in the availability of external information about the practices promoted or other practices competing with or complementing them. Examples of these include communication activities of others, such as the media, the government, nongovernmental institutions, or other agricultural chemical companies.
- emotional factors—that is, ongoing changes of the value system or changes in the affective environment of the farmers. The value system of a society is a dynamic entity. The values attributed to health and safety may have been in a process of change when the project began.

The basic methodology chosen to identify and quantify the effects of the project consisted of comparing the variables we were interested in before and after the programme and, simultaneously, with and without the project (in the intervention and control areas, that is). The changes observed in the test area were a blend of changes that would have occurred without the programme and those induced by it. Quantitative comparison of the changes in the intervention and control areas allows the identification of programme-induced changes, provided that the external factors are the same in the two areas. The greater the similarity between test and control areas, the more reason we have to assume that the changes in the control area indicate what would have occurred in the test area without the communication programme—an assumption required to identify the programme-induced changes in the endline KAP study.

The test and control areas therefore should be as similar as possible, especially with respect to characteristics that might influence any of the variables of interest, such as economic activities and their profitability, climatic variables that influence agricultural production, education level, non-project-related media active in the area, and the value systems of the population. At the same time, to avoid potential spillover of the messages communicated, the control area should be isolated from the communication programme, which requires a certain geographical distance as well as limitation of the geographical area covered by the media used in the test area. Yet this very geographical distance implies that some external variables will also vary between the two areas. Thus there is a trade-off between the spillover potential of the project and the diversity of the two areas. Researchers must try to figure out the best balance in this trade-off before implementation, and then fix the boundaries of the test and control area accordingly.

Another important aspect in the choice of areas is size. The area chosen must be large enough to ensure that the interviews carried out will be sufficiently diluted that they do not unduly influence the average KAP of the population. Yet having too large an area may make it more difficult to find a suitable control area, increase costs, or, if costs are to be the same, dilute the communication effects.

The project test and control areas were chosen with these considerations in mind to the satisfaction of the researchers. In retrospective, there were a few specific problems in the selection of areas for the project. In Mexico, the climates of the test and control areas were known to differ slightly, but we did not expect that the difference would vary with years or seasons. (Generally, similarity between the two areas is important, but even more important is the requirement that any difference that might exist should not vary during the intervention period.) In India, the fact that the control area surrounded the intervention area proved more of a problem than we expected in terms of hindering our ability to afford a spillover effect. Zimbabwe, on the other hand, did not present these problems and no significant spillover occurred between the selected areas.

In addition, several economic and information factors that could not have been anticipated may have affected the study's results. A drought in Zimbabwe, for example, changed farmers' perceptions of the importance of effectiveness versus safety. By the same token, an earthquake in Villaflores (the test area in Mexico) in October 1995 temporarily set back efforts to interest local farmers in the intervention programme. And in the control area in Mexico (Cintalapa), a communication programme by a competitor around the same time as the project may have led to many of the favourable changes in knowledge levels that we found there at the end of the study period.

In fact, it is virtually impossible to keep all the key factors in place in a control or test area during the entire time of a research project, which in the case of the project spanned more than five years. The scientific methods for this social research were developed in the late nineteenth and early twentieth centuries, when people did not move around so much or so often, and when there was a fraction of the information about societal change available to the general public that there is today. Indeed, the idea of a "control" area may belong to the past; new methods of research may be needed, given that long-term studies and appropriate control areas cannot be selected as they were in the past. Also, it is never easy to teach new behaviour during stressful times, such as exist in all three areas chosen for this research. Farmers are inclined to leave things the way they are, even when they know their practices could change for the better, simply because they have too many problems to cope with. In a time of constant internal migration, no part of a country can remain unchanged over five years.

Even without confounding variables, there were inevitably some differences between the intervention and control areas. While these were not large (although this is admittedly subject to the judgment of the researcher), statistical methods may nevertheless be used to identify the effects of the programme. Assume, for example, that there is a difference in education levels between the two areas. Since education potentially affects the reaction to changes in the economic, social, or affective environment, diverging educational levels between the two areas could cause diverging reactions to external influences. The assumption that the changes in the control area reflect those in the test area that would have occurred without the programme could then not be sustained. But since education varies within the control area as well as within the intervention area, it is possible to compare subpopulations of similar education levels in the two areas, and to assume that although the overall effect of external changes may have differed, the effects on groups with the same education did not. A number of such analyses of the data collected in the project were performed with logit adoption models; these are available upon request.

A further problem of identification refers to the adoption that would have happened in the absence of the programme. In the case of this project, we did not generally introduce completely new concepts. Rather, the promoted knowledge and the recommended practices were already present, to varying extents, before the intervention. Since we assume that the recommended practices are economically or otherwise beneficial to the farmers, we must also assume that they would slowly have been adopted anyway by more and more farmers, through some diffusion process due to communication between farmers. If the adoption curve is S-shaped, as commonly believed, we would have expected that this diffusion process would be relatively slow for practices hardly known at the time of the baseline, but that it would be faster for practices that had already been adopted by a considerable proportion of farmers at that time. In practice, this means that an increase in the adoption level from 10% to 20% over some given time period is likely to be more meaningful than an increase from 40% to 50% over the same period, even though the percentage point increase difference is 10 in both cases.

Choosing Suitable Variables

As noted earlier, the two basic objectives of the communication, education, and training programme were safety and effectiveness in the use of agricultural pesticides. But these are abstract concepts that cannot be measured directly. Thus surrogate variables had to be used to track down the effects of the programme.

Safety

Safe use of pesticides relates to the probability of an individual becoming ill due to exposure. Statistics—the science of calculating probabilities—provides the means of assessing this risk, provided the data are available. But there were some practical problems involved in assessing safety in the project. Certainly the most important of these was obtaining valid health data.

Health damage that might result from exposure to pesticides is classified into chronic and acute effects. Long-term health outcomes belong to a group of degenerative diseases that usually have many possible sources. There is thus no possibility of making automatic associations of health outcomes and their causes. This makes the investigation of long-term health risk factors inherently difficult (Maroni and Fait, 1993). A few more recent papers report statistically significant links between chronic health impairments or labour productivity and past pesticide use (Antle and Pingali, 1994; Antle, Cole, and Crissman, 1998).

In the case of the project, the chances for accurate estimates were further reduced by the fact that we wanted to assess differences (the difference in health effects with and without the programme). In order to estimate epidemiological health models, the researcher needs data on the health status of the population under study and any variables that may have influenced it. The assessment of health status presents its own methodological problems, which are discussed later. However, the more difficult part often is assessment of the factors that may have caused a health problem. In the case of chronic health damage, these may have occurred many years ago. This makes a reliable assessment very difficult. For example, in the case of pesticide-related chronic illness, the researcher would have to try to collect reliable information about the ways pesticides were used, the brands used, and so on over the past 10–20 years for every subject in the study sample. Obviously such information cannot be very precise. Researchers therefore try to use approximations. Whatever variables are used, however, they are only proxies for true occurrences in the past that may have caused the problem.

Unfortunately, the use of approximate explanatory variables has a serious impact on the statistical precision of the epidemiological model. This is the most important reason it has been impossible to even ascertain the existence of an effect in most cases investigated (Maroni and Fait, 1993), let alone to quantify the effect of a seven-year programme in a small population. Therefore, the Steering Committee decided early on not to make any attempts to estimate long-term health damages due to pesticides and the programme's possible effects on them.

The Committee did decide, however, to assess the programme's effects on acute health problems, which required an assessment of acute health damage. Official data on such problems are often deficient or lacking. In addition, the health data had to be related to the assessed KAP. We decided, therefore, to collect our own data.

The data could be collected in two ways: medical surveys complemented with interviews, or interviews alone. From a physiological point of view, medical surveys of farmers should yield quite precise data on health status. The monitoring might be carried out on farm operators after spraying, thus increasing the proportion of positive cases and reducing the data collection effort, or on a random sample, which would require a larger sample size. In either case, this type of survey would require having a team of physicians and/or nurses in the villages of the pilot areas for several days, asking farm operators for permission to do physical examinations. Clearly such an undertaking would lead to tremendous gossip and would affect awareness of health problems in the small villages in which we worked. As the research instrument thus would have confounded the effects of the communication programme, the idea was abandoned.

The second option—interview data—was expected to cause less researcher bias because interviewers have a much lower profile than medical doctors and because an interview creates less awareness than an interview plus a physical examination. But there are other problems associated with health data gained from interviews. First, awareness plays a role. Most acute health problems due to pesticide exposure are "minor" symptoms such as headache or giddiness—symptoms that are similar to those of other diseases commonly occurring in the project areas, such as malaria. The probability that a farmer remembers and reports such symptoms as having occurred after spraying or otherwise handling pesticides is expected to increase with growing awareness and knowledge relating to the health risks of exposure to pesticides. This might not be a big problem if we just wanted to assess the level of such symptoms at one point in time, but we wanted to compare the level before and after the programme. Since the programme promoted awareness of health, the percentage of these minor health effects reported at the interviews was expected to increase in the test area over the intervention period. There might even be an increase in the control area, due to the awareness created by the baseline survey and by possible spillover and diffusion effects.

Farmers are more likely to remember severe acute health damage, however, even before awareness is enhanced through the programme and the surveys. This would tend to make data on severe acute health effects more reliable than those on milder effects. On the other hand, severe effects are also much less frequent, which reduces the statistical reliability of their estimate. In conclusion, the acute health effects—both minor and severe—reported in interviews did not constitute a very reliable data source for our purposes either.

A second problem that was often discussed by the Steering Committee may not have been as worrying as it first seemed. We were concerned that respondents might not be able to distinguish pesticide-related health damage from other illnesses, especially because most acute symptoms of pesticide exposure are nonspecific. The solution to this problem lay in asking respondents not about health problems "due to pesticides" but in general about their health state during or after a day of pesticide spraying. The linkage to the communication programme in the intervention areas would be merely statistical. In the case of the question used to elicit health problems ever suffered due to pesticides, it was impossible not to mention the reason in the question (that is, to ask for problems due to pesticides). But, as mentioned, the main problem associated with questions about health status is that of the awareness created through the programme.

Given that neither chronic nor acute health states could be assessed very reliably, surrogate variables were used to measure the achievement of the safety goal. Basically, this meant assessment of the adoption of safe practices. Awareness may also influence reporting of such practices, but this problem was remedied by complementing the interviews with observation surveys.

It has been noted that observation surveys are subject to research bias if the farmers are aware of being observed. The researcher bias argument states that the visible presence of an observer might induce farmers to change practices temporarily in favour of those they thought were expected. This bias might thus also increase with awareness of the programme. It was not possible to observe the farmers from hiding, however, because some items to be observed required that the observer be on the spot where the pesticides were mixed. Moreover, the observers had to contact the farmers before observation because some of the more remote fields could not be found easily. Farmers were therefore aware in advance that they would be observed.

Starting from the hypothesis that farmers would gradually return to usual practices as they got used to the observer's presence, we tested for the existence of such a bias in Mexico's 1995 observation survey. The farmers were observed during a whole workday each, so they were generally spraying between 3 and 10 tank loads of pesticides. They would generally use the same practices from the first to the last tank load; when they changed their behaviour (such as putting on or taking off a garment), there was no clear direction towards the better or the worse practice. The existence of any bias due to the presence of the observer could thus not be confirmed in Mexico. (Although the same could not be said for Zimbabwe.) In conclusion, we believe that the observation data on practices in Mexico are in general representative of true practices. These data are complemented by the KAP surveys, which deal with unobservable concepts and were carried out on larger samples.

Thus, at this stage, we measured success of the safety part of the programme by looking at the adoption of safe practices. Of course, this method is valid only if we know which practices are safer than others. There is a large body of literature that lists the practices that reduce pesticide exposure under a wide variety of circumstances. But the exposure trials that such data are based on were, to our knowledge, always conducted under what researchers considered to be "reasonable use" conditions. Often the objective of such trials was the assessment of the potential for protection, which may differ significantly from the actual protection under field conditions. Other trials imitated field conditions but, for reasons of scientific scrutiny, focused on only a few elements (such as investigating the effects on exposure of a garment, but not of that exposure in combination with a leaking sprayer).

In addition to this information on experimental and semi-experimental trials, there is literature about practical experiences. This allows some firm conclusions about the protective effect of some practices (such as hygiene measures). But it also shows that protective gear is not necessarily used well and that, if used badly, such gear may actually expose the operator or others to the spraying mixture. This is why the assessment of practices such as the use of protective gear does not give the full picture. It should be complemented with other information that tells whether the practice thought to be safe was indeed safe. This information may consist of nonquantitative experiences documented by the extension workers of the programme or complementary variables assessed in the KAP or observation surveys.

Another way to close the gap between practices and health status is the estimation of health risk models. These relate reported or observed practices to actual health states, thus corroborating or rejecting the assumptions regarding the safety of certain practices. Such models are estimated preferably with cross-sectional data (either baseline or endline, but not both), due to the effect of awareness on the reporting of ill health. They also give quantitative information about the reduction of health risk achieved by given practices, and thus help to interpret the quality of the induced adoption of safe practices.

In summary, the advantages and drawbacks of the different variables that might be used to estimate the achievement of the safety objective are as follows:

- Chronic health variables are impossible to assess with sufficient accuracy to estimate the effect of a programme, especially in pilot areas limited to some thousands of farmers.
- Medically assessed acute health state would cause severe researcher bias.

- Interview-assessed acute health variables are subject to awareness bias.
- Practices found to be safe in the literature are fine for those that are really safe under all circumstances, but problematic for others that may be unsafe if not implemented correctly.

Effectiveness

The second objective of the programme—effectiveness in the use of pesticides—poses quite similar measurement challenges. Effectiveness of a pesticide may be defined as the percentage of a pest population killed, but ultimately the objective is to improve the profitability of the agricultural operation. The theoretically most appealing way to assess profitability therefore involves an estimation of a production model, including a production function and a model for farmers' decisions regarding input allocation. This has been done by Angehrn (1996) for the Indian test area, where he found that scouting cotton pests allowed a reduction of pesticide use of one fourth (by value) while achieving the same level of control. However, Angehrn also pointed out some drawbacks of common production models that include pest control, which are related to the required simplification of pest control strategy variables and the lack of data on pest populations and pressure at the start of the season.

These drawbacks make it desirable for our evaluation of the programme to include the adoption of practices that are presumed to be more effective than others. As in the case of safe practices, the certainty with which the relative effectiveness of pesticide use practices is confirmed may vary, and should be subject to careful consideration.

Overall

In conclusion, the evaluation of the effectiveness and safety dimensions of the programme both require the use of indicators complemented by a discussion and, where possible, statistical confirmation of the suitability of the indicators.

Model

Another problem arises if there are differences identified during the baseline KAP study in the use of some safety practice. Say, for instance, that the use of rubber boots during the spraying of herbicides stood at 20% in the test area before the project began and at 5% in the control area. Then assume that by the endline KAP study the level in the control area increased to 15%. If the effects of external factors over time in the two areas were similar, what percentage of farmers in the test area would wear rubber boots if there had been no intervention? Would it have been 30% (20% + 10%), since the control area increase was 10%? Or is it equally logical to assume that it would have been 60%, since the control area level tripled? This illustrates the problem of the mathematical form of adoption. The first figure was arrived at by assuming that the effect of an external change on adoption of this practice was additive, while the second one assumed a multiplicative model.

Both the additive and multiplicative models may be valid in some ranges of adoption. In particular, additive models seem to be suitable for adoption levels near 50% and for minor changes in adoption, while multiplicative models are suitable for initial levels near 0% as well as for minor changes in adoption. But neither model may be suitable in all cases. Assume for a moment that the baseline level in the test area in the example just described had been 50% and that the control area level increased from 20% to 80% during the study. The hypothetical endline test area level would be estimated at 110% (50% + 60%) using the additive model and at 200% (50% quadrupled) using the multiplicative model. Neither of these conclusions is sensible, of course.

Other mathematical models do keep estimated adoption levels in the meaningful range between zero and 100%. They are typically S-shaped, meaning that the rate of adoption of a new technique or practice is initially low (flat adoption curve), accelerates as the technique becomes more common (steeper slope of the adoption curve), and tends to slow down again as the level of adoption reaches some maximum.

In order to avoid the disadvantages of comparing proportions in terms of differences, S-shaped mathematical models are widely used in adoption studies. An example of an S-shaped adoption curve is the logit model. In the project, logit models were estimated for a selection of variables in each country; as noted earlier, these are available upon request. For simplicity's sake, the main results presented in Chapters 3, 4, and 5 use additive models.

Overall, the logit models showed that the difference between the KAP changes found in the test and the control areas—that is, the "net change"—on its own is not necessarily an efficient measure for the incremental adoption caused by the project. Especially where major changes occurred in the control area, the direction of the net change may be opposite to the direction of the estimated effect of the main media. Apparently, the assumption that outside factors influence both areas to similar extents is easily violated, given the considerable amount of time that passed between the baseline and the endline surveys.

The logit models also showed that the project improved knowledge and practices related to pesticide safety in all countries (as detailed in the chapters that follow). They suggest that these improvements translated into a reduction of acute health damage due to pesticides, although strong direct evidence for this result was not established when all the data from different surveys were considered.

Aggregation Level

At the most general level, we were interested in linking the programme as a whole to the benefits it produced as a whole, which raised aggregation problems. At a more detailed level, we were interested in which elements of the programme were particularly successful and which failed, which raised the problem of disaggregation.

As mentioned in previous sections, a disaggregation of the programme objective into indicators for safety and effectiveness was required to circumvent some of the difficulties involved in directly assessing health risks or the profitability of the crop. This also told us which indicators reacted most to the programme. Thus the evaluation of achievements was carried out at various aggregation levels.

On the side of the causal factors, we also can operate at various aggregation levels. At the most general level, the impacts of the programme as a whole had to be evaluated. Beyond that, it was of interest to assess the reach of different elements of the programme—that is, different recommendations and communications media. This analysis could be done only partly with statistical tools and had to include qualitative, impressionistic information from all the researchers and extension workers.

Statistical tools may indicate, for example, the reach of each media. But the interpretation of the results remains a question of the sensibility of the researcher. This is because of the numerous interactions between the variables involved in the adoption process. For example, a common impression of the project collaborators was that safety messages would only be accepted if brought forward by some respected person, and that the extension workers could gain this respect by helping farmers improve the effectiveness of pesticide use or, more generally, the profitability of the crop. Another example of possible interactions involves the shift in practices that follows improved awareness and affective changes, which in turn is reinforced by the experiences made possible by the new practice.

Cost-Benefit Analysis

A cost-benefit analysis (CBA) makes the achievements of a project comparable to its costs and to projects of a different nature by expressing them on a common, monetary scale. In theory, such a comparison is a formidable base for deciding whether to implement a project at all or whether to implement one project versus another one. In practice, however, the usefulness depends essentially on the quality of the information used to perform the CBA. Such analyses in general are delicate tasks because they usually require the researcher to make many assumptions that complement the information elicited on site. In the case of this project, the usual challenges of a CBA were multiplied by the fact that the primary data itself were subject to the methodological problems described earlier.

In view of this, the main lesson learned in the CBA attempt in this case refers to the methodological problems described below. The actual results do not contribute to the discussion as much as we would have hoped. (The full study is available on request.)

A CBA has three stages:

- identification of benefits and costs,
- measurement of benefits in physical units, and
- valuation of benefits and costs (Warner and Luce, 1982).

Each of these steps presents its methodological challenges.

Identification of Benefits and Costs

The objective of the CBA is to provide information to decisionmakers confronted with the question of whether to implement an information and education campaign to make pesticide usage safer and more effective, ultimately aiming at an increase of the farmers' welfare. The identification of benefits and costs follows this premise. Thus, the programme benefits and costs to take into account are those that could be expected from a similar, future programme.

Benefits namely include improvements of health. These benefits are understood net of costs incurred by the farmer to achieve better health. For example, the net health benefit is the improvement in health minus the incremental cost of protective gear, the opportunity cost of the additional time spent washing the clothes and maintaining the equipment, and so on. Identification of these benefits is subject to all the uncertainties arising from the methodological problems discussed earlier, such as potential spill-over, external influences, differences between the test and control areas, and so on. Therefore the CBA could not be done without returning to a considerable number of assumptions and scenarios, which gave interesting but rather theoretical insights.

Measurement of Benefits

The primary project goal, improvement of farmers' health, could not be measured directly because reported health turned out to be a blend of the underlying physical reality and the respondents' attitudes towards health. This latter concept changed, however, during the intervention period and therefore eluded analysis.

As far as the physical description of project achievements is concerned, we turned to indicators of safe use in order to draw conclusions about health. That is, instead of stating that health has improved by so much, we are able to describe the adoption of safer practices, which should have produced a health improvement. For a CBA, this procedure is not sufficient, because that analysis needs to put monetary figures on the improvements. This is a challenging task even if the health improvements are known precisely, but it is virtually impossible if only indicators of safe use are available.

This problem was recognized early in the project. We anticipated that the communication campaign would increase farmers' awareness of pesticide-related ill health and incite them to report milder problems over time. We therefore designed additional surveys (referred to as CBA surveys) to assess the severity of health problems in addition to their incidence, which was already elicited in the KAP surveys. As a measure of severity, we used the resting period that followed incidents of suspected intoxication with pesticides. The result of those surveys was that indeed the average severity of reported health problems tended to decrease between the baseline and the endline, but the reliability of these data was hardly sufficient to support the analysis.

A second problem involves the potential chronic health effects. As for the analysis of achievements in physical terms, it is permissible to completely abstract from such effects. After all, decreasing acute problems are an indicator of reduced pesticide exposure, which in turn would lead to the expectation that any chronic health problems that might exist would at least not increase. By contrast, when health benefits are to be valued for the purpose of comparing them with project costs, long-term health effects should be included as well. Measurement of these effects was not possible, however.

A third problem of measurement refers to the need to estimate benefits that arise in the future. The endline and follow-up surveys give some indication of the persistence of improved practices, but unfortunately do not eliminate the need to make important assumptions regarding the future. Crop protection changes and the knowledge conveyed may become inadequate or obsolete with time. The speed with which this occurs depends, among other things, on progress made with the development of new, less toxic pesticides and other crop protection technologies, and on pesticide regulations. These factors are difficult to predict, implying that the future benefits of the programme are subject to considerable uncertainty.

Valuation of Benefits

Under economic theory, the value of a good is not an intrinsic property but the value that individuals attribute to it. For an individual who does not currently have the good, this is the amount of money the person would be willing to pay for it (willingness to pay, WTP). For an individual who already has the good, it is the amount of money the person would accept by means of compensation to give up the good (willingness to accept, WTA). Thus, the value or cost of a health risk is the WTP to avoid the risk of a person subject to it, or the willingness to accept the risk of a person who is not currently exposed to it. For traded goods, the market prices are approximations to the WTP and WTA. Health as such (as opposed to health care or treatment), however, cannot be bought on a market. This makes its valuation more challenging.

Valuation methods based on the WTP and WTA concepts are categorized as direct and indirect. Indirect methods analyse behaviour of the population of interest and try to infer the underlying values. For example, a researcher observing that a farmer incurs a health risk by using pesticides would conclude that the cost that the farmer attributes to the health risk is less than the cost of any alternative that would eliminate the risk. One of these alternatives is not using pesticides. This would result in a profit reduction. By this logic, the fact that farmers use pesticides is in itself an indication that health risks or costs are lower than the productivity benefits of pesticide use. Behaviour (pesticide usage in this example) reveals preferences, which is why indirect methods are also called revealed preference methods.

Direct methods, in contrast, confront individuals with hypothetical choices. They are examples of stated preference techniques in that individuals do not actually make any behavioural changes, they only state that they would behave in a certain fashion (Adamowicz, Louviere, and Williams, 1994). The most prominent direct method is the contingent valuation method (CVM). Individuals are asked how much they would be willing to pay, for example, for a given health improvement. We applied the CVM to health risks due to pesticides, asking farmers about their willingness to pay a higher price for hypothetical pesticides that the interviewers described as completely inoffensive to human health but equally effective as existing products currently used by the farmers. (See Angehrn, 1996, where the CVM applied to India is presented.)

The CVM poses several peculiar and most interesting methodological challenges (Diamond and Hausmann, 1994). The most important one, in our context, might be the fact that it is based—as is economic theory—on individuals' preferences. When preferences are what we tried to influence through the communication campaign, how can we then use it as a basis for valuation of health benefits? Moreover, the CVM requires that the respondents be "sufficiently familiar" with the benefit to be valued. Although we found that farmers are familiar with acute intoxication symptoms, they are not aware of ways to prevent these problems and thus are not familiar with the benefits of prevention. Neither can they be expected to appreciate chronic health risks qualitatively, let alone quantitatively. In conclusion, the CVM probably is the only—but not a particularly satisfying—method to put monetary values on health risks due to pesticides and their prevention.

Taken together, these limitations led to the conclusion that the CBA results were not sufficiently reliable to make a valid and practical contribution to the discussion and interpretation of this project's results.

Bibliography

Adamowicz, W., Louviere, J. and Williams, M. (1994) Combining revealed and stated preference methods for valuing environmental amenities. *Journal of Environmental Economics and Management* 26: 271–292.

Angehrn, B. (1996) *Plant Protection Agents in Developing Country Agriculture: Empirical Evidence and Methodological Aspects of Productivity and User Safety*. Thesis No. 11938, Swiss Federal Institute of Technology, Zurich.

Antle, J.M. and Pingali, P.L. (1994) Pesticides, productivity, and farmer health: a Philippine case study. *American Journal of Agricultural Economics* 76 (August): 418–430.

Antle, J.M., Cole, D.C. and Crissman, Ch.C. (1998) Further evidence of pesticides, productivity and farmer health: potato production in Ecuador. *Agricultural Economics* 18: 199–207.

Diamond, P.A. and Hausmann, J.A. (1994) Contingent valuation: is some number better than no number? *Journal of Economic Perspectives* 8 (4): 45–64.

Feder, G., Just, R.E. and Zilberman, D. (1985) Adoption of agricultural innovations in developing countries: a survey. *Economic Development and Cultural Change* 33: 255–298.

Maroni, M. and Fait, A. (1993) *Health Effects in Man from Long-Term Exposure to Pesticides: a Review of the 1975–1991 Literature*. Elsevier, Amsterdam.

Rogers, E. (1962) *Diffusion of Innovations*. Free Press of Glencoe, New York.

Warner, K.E. and Luce, B.R. (1982) *Cost-Benefit and Cost-Effectiveness Analysis in Health Care*. Health Administration Press, Ann Arbor, Michigan.

3

Mexico

Overview of the Agricultural Sector

Because of topography, climatic conditions in Mexico vary considerably, a large proportion of the country being dry. One peculiarity of Mexico's geography is that there are few large rivers, and water resources are unevenly distributed. Topography and climate make only about 21% of the country suitable for arable farming and a further 57% suitable for pasture. Forests and woodland cover around 17% of the land.

Although northern parts of Mexico are very dry, suffering drought for 5 or 6 years out of 10, they have become the most productive areas for agriculture. This is a result not only of irrigation schemes, but also the establishment of large privately owned farms in the area. Farms in the centre and south have lagged behind, with their main problem being the landownership system. As a result of agrarian reform (intensive in the 1930s), these farms are predominantly *ejidos*: land allocated by the government to individuals or communes that would not be sold, rented, or mortgaged, but just inherited and subject to expropriation if not worked upon. The system caused increasing fragmentation of farming units, insecurity, and greater inefficiency. The Salinas government amended the constitution in 1992, paving the way for the *ejidatarios* (owners of *ejidos*) to be given full property rights to 4.6 million plots on the 30,000 *ejidos* that account for half the country's area. However, the process of distributing titles has been slow.

The Salinas government also privatized sugar mills, the tobacco company, and parts of the national basic foods supply company, Conasupo. In 1993 a new scheme, Procampo, was introduced to replace the former system of price support

---

This chapter is based on material prepared by Beda Angehrn and Jean-Paul Serres except for "Overview of the Agricultural Sector", which is excerpted with permission from Economist Intelligence Unit, Mexico: Country Profile 1998–99 *(London: 1998), with minor stylistic alterations.*

for basic grains, providing yearly cash payments for 15 years to producers of cotton, rice, barley, beans, maize, sorghum, soybeans, and wheat.

Inadequate investment and low productivity continue to affect the sector, exacerbating the extreme poverty in rural areas. Its growth has consistently underperformed that of the rest of the economy (growing by 1.4% in 1997), and as a proportion of total gross domestic product it fell from 5.8% to 5.2% between 1993 and 1997. At the same time Mexico has tended to run deficits in its foreign trade in agricultural products. In 1992–97 only 1995 registered a surplus, because of recession and high prices. During 1997 the deficit was a relatively low $344 million. In 1995 President Zedillo initiated his *Alianza para el Campo* (Alliance for the Countryside) programme, which reinforces the Procampo system of direct cash subsidies by providing that these should be maintained in real terms for 15 years.

To support its agricultural programme, the government has raised the level of financing per hectare and is improving the flow of credit from state agencies. In 1996 it also had to step in with a rescheduling plan for debtors in the farming and fishing industries as well as providing emergency funding to help the sector confront the third consecutive year of drought. By October 1997, 75% of all debtors qualifying for this programme had adhered to it.

Despite the sector's problems, Mexican crop production is among the highest in the world. Overall it accounts for more than 50% of sectoral output, with maize, beans, wheat, and sorghum being important for the domestic market while coffee, sugar, fruit, and vegetables are the leading exports. Maize (with 8.1 million hectares), sorghum (2.1 million hectares), and beans (2 million hectares) occupied by far most of the 16.9 million hectares sowed during 1998. The livestock sector contributes about 30% of sectoral output. During 1997 meat production, the most important activity, increased by 6.1%, to a record 3.8 million tons; milk production increased by 3.3%, to nearly 8 billion litres, also a record.

Target Area

The project Steering Committee decided to undertake its research and communication campaign in Mexico in the state of Chiapas, bordering on Guatemala in the southernmost corner of the country, where nearly 23% of the country's water resources are located. Chiapas is characterized by extreme social discrepancies. Land reform arrived there comparatively late. In the late 1920s, the old Chiapas families who had dominated politics during the previous 50 years were defeated for the first time by a coalition between a peasant leader and a reform-friendly landowner and army general. After several setbacks, these reformers gained power as President Cárdenas enacted profound land reform. Cárdenas's governor in Chiapas sought to mobilize Indian migrating labourers, coffee pickers, and landless peasants to gain support for the new regime and the reform project. They succeeded in breaking with enslavement, enforcing a minimum wage and cash payment of rural labourers.

Land reform transformed many large coffee-*fincas* (farms) of the Pacific coast into *ejidos*. The Indians of the highland communities around San Cristóbal got back many of the lands they had lost to the conservative *hacendados* in the nineteenth century. But properties of prerevolutionary size and dimensions persisted in some communities. Along with the reforms, government officials started to integrate the newly founded peasant movements and unions into a corporate political system that allowed them to conserve their political power, combining and balancing selective participation and control. After a phase of increased power for farmers in the late 1940s, the political orientation in the capital changed and an oligarchy of cattle breeders and *finqueros* transformed the organizations from institutions that balanced political interests into instruments of political control over the rural population. They continued expanding their properties, cut forests in the Lacandon jungle, or converted into pasture the land they once had leased to highland Indians.

This is the story behind the social and political structure in Chiapas today. Conflicts are most acute in communities where land reform had little impact as well as in the Lacandon jungle to the east of the highlands, where the interests of expanding cattle growers and the population pressure of the Indians collided most heavily. During the economic crises of the 1970s and 1980s, opposition increased. In some cases, Indian peasants occupied land that was legally owned by *finqueros* and *hacendados*. Dozens of more or less radical peasant movements arose in Chiapas during the 1970s and 1980s. Around San Cristóbal, opposition took the form of conversion to Protestantism. Given the cultural and social conditions, this may be interpreted as protest against the old order, with its *caciques* (local power brokers). Although not unique to Chiapas, these events prepared the soil for the rebellion of the Zapatista National Liberation Army.

Within Chiapas, the Fraylesca is politically and socially the most stable region, although there are occasional land occupations, especially in the hilly coffee-growing areas. The Fraylesca is prosperous due to a fairly well developed infrastructure, relative certainty about property rights, and the associated productivity of agriculture and the commercial sector. The population is ethnically quite homogenous, and has access to schools and health care. The conflict in the highlands and the Lacandon jungle therefore affected the project area of Villaflores only indirectly.

The overwhelming importance of maize for much of the social sector was an important reason for choosing Villaflores as the test area for intervention. Villaflores is one of four municipalities of the Fraylesca region, an area sometimes called the "Chiapas corn belt". (The central town in the municipality is also called Villaflores, and so is one of the largest *ejidos* in the test area.) It is in the southwest of Chiapas, at 600 metres above sea level and separated from the Pacific coast by the Sierra Madre de Chiapas. Villaflores covers 1,232 square kilometres and had a 1990 population of 73,311 inhabitants. The test area consists of 24 major villages in Villaflores and neighbouring Villacorzo, with some 11,000 adults residing in 3,300 farm households. The area has 25–30 private doctors, one public hospital, and approximately five public health centres.

The terrain is generally hilly, with some plains along the major rivers. The climate is characterized as tropical subhumid with summer rains. The rainy season is from mid-May to October, with 1,209 millimetres average precipitation per year.

The mid-summer dry period of one to several weeks, known as *canícula*, may cause water stress for maize, in particular where the organic matter content of the soil is low due to the traditional practice of burning stubble before sowing. This custom has required a special permit since 1992 and is now slowly being abandoned. *Ejidatarios* grow mainly beans and sorghum in addition to maize. The fields on river banks are often irrigated and planted with vegetables in the winter season. However, only a very small percentage of the *ejidatarios* have at their disposal water and pumps for irrigation. The farmers who have traditionally held full property rights to their lands (*pequeños propietarios*) grow maize and sorghum and keep dairy cattle. Major agricultural problems are soil acidification, insect pests of maize and beans, perennial maize weeds, and soil erosion on steep slopes.

Once Villaflores and Villacorzo had been proposed as the test area, a similar but sufficiently remote control area had to be found. Other areas within the Fraylesca were not appropriate because economic and cultural exchanges among the *ejidos* and municipalities of this area are intense. The Cintalapa region was chosen; it consists of 10 *ejidos* of the municipalities Cintalapa and Jiquipilas, located to the northeast of Villaflores, in the eastern extreme of Chiapas.

The towns of Cintalapa and Villaflores are separated from each other by a hilly range of pine forests, where agriculture is sparse, which is crossed only by an extremely windy and rough road of about 110 kilometres. The distance between the closest *ejidos* of the test and control area is about 70 kilometres. Like the Fraylesca, the Cintalapa area is separated from the Pacific coast by the Sierra Madre de Chiapas, although the mountains are not as high in this region. To the north, Cintalapa is bordered by a vast wilderness area called Chimalapas Region.

The town of Cintalapa is a little smaller than Villaflores. It is well connected to the state capital Tuxtla Gutierrez by the Panamerican Highway. This, and the fact that its area of influence is smaller than the Fraylesca, may explain why Cintalapa is a less important commercial centre than Villaflores.

Cintalapa's agricultural structure is a little different from that of the Fraylesca, with a larger proportion of private landholdings, more cattle farming, but no coffee. Our focus was on the *ejidal* maize economy, however, where the differences were less pronounced. Cintalapa is not as famous for its maize production as Villaflores, and project managers assumed that the *ejidatarios* of Cintalapa derive less of their identity from being maize producers than their colleagues in Villaflores would. Since the end of the research portion of the project in 1997, a large proportion of former maize farmers started to grow sorghum, due to a shift in grain prices. Rainfalls are somewhat more erratic in Cintalapa, a fact that confounds the analysis of maize yields.

Baseline KAP Study

Initial surveys of farmers' knowledge, attitudes, and practices (KAP) regarding pesticides were conducted in August and September 1992. The target audience for the study consisted of *ejidatarios* rather than traditional private landowners. The interviews were done in Spanish by a market investigation firm under the supervision of the marketing department of Novartis Mexico. In both Cintalapa and Villaflores—the control area and the intervention area—500 small-scale farmers were interviewed about their use of pesticides.

A qualitative survey was done that consisted of focus group sessions, which typically included 8–10 farmers and were guided by a motivational psychologist. To better capture the opinions of different segments of the target population, the groups were differentiated by the participants' age, their pesticide intoxication history, and, at the completion of the project in the test area, their attendance at project meetings. During the baseline study, focus group sessions were also conducted with pesticide dealers.

The baseline KAP survey established that the average target farmer had a primary school education and cash household expenses of about US$6 a day. Some of those with a primary school education were not able to read well, however, and 13% had no formal education at all, so the functional illiteracy rate was about 30%. A television set is available in virtually every household (92%), a radio in 78%, and a refrigerator and a bicycle in about half the households.

Four fifths of the target farmers planted maize only. Two out of five reported that they reuse their own seed without ever buying new seed. The remainder bought certified seeds, but usually reused the harvested seed at least once. Some farmers also cropped beans, peanuts, or vegetables. Insect pests and weeds were considered the major problem of agricultural production. Second in importance were water problems (either too much or too little). Information on farming was obtained from fellow farmers (65%), from governmental technicians (50%), and from input dealers (9%). Fellow farmers were considered the most reliable information source.

Ninety per cent of the farmers affirmed that a health risk is associated with pesticide use, and 25% reported that they had at some point suffered a health problem due to pesticides. Most correctly knew that the operator is the person most at risk, and more than half affirmed that everyone who enters the field during spraying may be at risk. When asked about the importance of the pesticide's cost, effectiveness, and safety, 63% of the farmers ranked safety first. This stands in contrast with what farmers reportedly inquire about when buying a pesticide: the "strength" of the product (46%), the price (31%), the recommended rate (30%), the control spectrum (25%), and the expiration date (17%). Only 12% of farmers knew that the toxicity is indicated by a coloured stripe, and 22% reported that they inquired about the product's toxicity. Pesticides were commonly bought just one day in advance of being used, and stored in a shed near the house.

Based mainly on the baseline observation and KAP surveys in the test area, the spray routine can be described. During the days prior to spraying, three quarters of the farmers had inspected the fields and observed insect pests or weed infestation. Before the work, which typically starts between 6:00 and 7:00 a.m., most farmers had a coffee and about half of them had some food at home. The houses of an *ejido* are situated closely together, forming a village. As a result, the fields may be several kilometres away. They were reached on foot, bicycle, ox cart, or horseback in 40 minutes on average.

Around 10:00 a.m. the spray work was interrupted by two thirds of the farmers for a more thorough breakfast, consisting of a maize-cocoa drink called "pozol" or some other food. Forty per cent of the farmers reported that they wash their hands before eating during breaks, which is particularly worrying because the pozol is sometimes stirred with bare hands. The spraywork was commonly terminated around noon.

The pesticides and the sprayer were mostly stored at home and carried to the field each day. About one fifth of the farmers had repacked the pesticide into a container other than the original, probably because the quantity left in the original container was more than they would use that day. Around 70% of the farmers do their spray job alone; the remainder are assisted by a family member or, in 10% of the cases, by a paid labourer. Half of the times an assistant was present the person was a minor.

In about two thirds of the observed cases, the air was still; farmers tend to stop spraying when it becomes windy in the afternoon. After that, 10% of the farmers clean or rinse the sprayer, throwing the rinsing water in the field. About two thirds of the farmers went home directly thereafter. The pesticide leftover is normally taken back home. Some 70% of the farmers washed their body partly or fully after spraying, either in a river or at home.

The pesticide-water mixture was commonly prepared directly in the tank of a common manual knapsack sprayer, which is deposited on the ground for that purpose. The water used for spraying was taken from creeks, rivers, puddles, or sometimes—although rarely—wells or the tap. If muddy, it was strained through a piece of cloth in addition to the strainer integrated in the sprayer. In 94% of the cases the pesticide was measured with an empty chili can; in other cases the container cap was used for that purpose. Neither of these tools has a handle, making exposure of the fingers to the pesticide concentrate relatively likely. About 8% came visibly into contact with the pesticide during mixing, once or several times during the day. The mixture was agitated by shaking the sprayer, stirring with the lance, or, in 56% of the cases, by alternately pouring water and pesticides into the tank. In 22% of the cases, the knapsack visibly leaked, mostly from the tap, the pumping mechanism, or the valve. The observers also reported visible drenching with the spray solution of the clothes (20%), the skin (24%), and the hands (48%).

The knapsack sprayers have a capacity of 15–20 litres and are equipped with a front lance with a single nozzle. This equipment requires farmers to pass once through each row of maize. Thus farmers walk up and down the field, spraying row by row. This implies that on windy days, farmers spray alternately with and against the wind, or with constant crosswind. On steep slopes, which account for about one third of the fields, maize is not necessarily planted in rows, and spraying patterns are more erratic. On occasions, farmers spray only parts of a field, where weed or pest infestation was observed to be particularly heavy. More than 80% of the farmers use hollow cone nozzles, although flat fan nozzles would be more adequate for most herbicide applications. Between one and two days of labour are required to cover one hectare, and each hectare requires between 8 and 12 tankloads. The average number of tankloads applied per day was 5. The most common pesticides were herbicide brands containing paraquat or 2,4-D, and insecticides containing methyl-parathion or pyrethroids. On about 12% of the applications, more than one brand was mixed.

Most farmers did not use any specific protective gear. The attire worn while spraying typically was identical to the clothes worn for other farm work. More than half the farmers wear sandals or no footwear at all when spraying, causing exposure of the feet to pesticides, especially because herbicides—the most prominent type of pesticides used in the area—are sprayed downwards to the weed or the soil. The remainder wore rubber boots (which provide significant protection) or shoes (which may provide some protection, depending on the material they are made of).

All farmers used long trousers, but some rolled them up, especially if the ground was wet, exposing their lower legs to the spray mist and the mixture sprayed to the weed. Virtually all farmers also used long-sleeved shirts, but again, many rolled up the sleeves. The usage of gloves was practically zero, despite the fact that one chemical company distributed thin plastic gloves together with one of the most commonly used herbicide brands. About 12% of the farmers used a handkerchief over their lower face, in an attempt to protect themselves both from pesticides and from the sharp maize leaves when spraying a large crop. The observation survey also showed that work clothes are used for several days in a row before being washed.

These practices contrast with the knowledge expressed in the KAP survey, where 33–39% of the farmers mentioned that gloves should be worn when mixing or spraying the most toxic pesticides, and 41% said that a mouth cover should be used. This illustrates the belief that the health hazard relates mainly to inhalation. In reality, dermal exposure accounts for the main health risks.

Cultural Characteristics of Target and Control Populations

The Fraylesca region has a low percentage of indigenous people (with the exception of a migrating group of coffee pickers), and therefore is ethnically more homogenous than the Chiapas highlands. Spanish is the common, almost exclusive language. Most inhabitants follow Catholicism, which is less influenced by the indigenous religion in this part of the state than in other areas. Television and radio are the predominant mass media, and daily newspapers are read by only a few people.

Until a few decades ago, the Fraylesca was quite isolated from the rest of the state. Surplus maize was transported on ox carts to Arriaga on the Pacific coast, a trip that took several days. The feudal system that prevailed before the actual implementation of land reform, and which is well remembered by elders, persisted for some time in different forms. Local power brokers continued to exhibit economic and political power, based on their relationship with the ruling party. The official institutions gained strength only slowly.

Today, there is an intensive cultural interchange with the state capital and the rest of the country though mass media and enhanced traveling possibilities. Migration to the United States, a dominant feature of many rural communities in central and northern Mexico, is much less pronounced in Chiapas; instead, emigration is predominantly to Mexico's own urban centres and, as far as the highlands are concerned, to former wilderness areas and petroleum extraction sites north of Chiapas. The area also enjoys immigration of Guatemalan labourers who work as coffee and plantain pickers.

Being an *ejidatario* was still an important factor of public life when the project started. The *ejidos* were part of the corporate political system. They, rather than individuals, received the government credits and subsidies before the recent deregulation. As a result, the *ejidal* commissioner (president) enjoyed substantial local power. Following the deregulation of agriculture and the abolition of the inalienability of *ejidal* lands, the *ejido* lost much of its significance. Today farmers are encouraged to form various types of local groups in order to get access to commercial and the remaining governmental credits and political power. These groups are smaller than and do not normally coincide with the former *ejido*, leading to a profound redistribution of power and pluralism within the community. The loss of power of the formerly predominant PRI also contributes to this evolution in the agricultural as well as nonagricultural sectors. Today's *ejidatario* derives his identity mainly from being a farmer, belonging to some civil group, and, to a lesser extent, belonging to a religious community.

The maize economy causes labour peaks during the vegetation period, especially during sowing and harvesting, but it also leaves the farm family underemployed during much of the year. Winter crops are planted only by the small group who have access to irrigation or river bank fields, and cattle breeding is still not a very prominent activity of the *ejidatarios*. Thus temporary off-farm jobs are in high demand; unfortunately, the supply is sparse.

Intervention Programme

The intervention programme was scheduled to begin in Villaflores in 1993, but it was delayed in order to conduct another baseline study with focus groups in the light of changes in the guidelines following information gained during research in 1992. The programme then actually began in 1994. (See Table 3.1.)

Maize is planted in Villaflores from mid-May to mid-July, and then harvested from October to December. The average yield is 3.7 tonnes per hectare (with just over half of the farmers getting yields of 3 tonnes per hectare). On average, each family sows 7.7 hectares of maize and produces 29 tonnes, but nearly half the families plant just 3–5 hectares and 27% plant 10–40 hectares. Private farms (*pequeños propietarios*) account for 40% of the planted area, while the social sector (*ejidatarios*) hold the other 60%.

During most of the intervention programme, six staff assisted with the communications campaign. (See Table 3.2.) The four male field technicians were agronomists who had just completed their studies. The field coordinator was a young man with two years of practical experience in the research department of Novartis Mexico. The female agronomist was a teacher at the local agricultural university.

The technicians had four weeks of training at the beginning of their employment in 1993. This included a didactics course, which was complemented and refreshed periodically at the beginning of each season. Due to the delay in the start of the actual intervention phase, the technicians had ample opportunity for practical experience of pest control in maize during the summer of 1993 and in other crops during the following winter season.

Based on the baseline KAP and observation surveys, the following key messages were suggested in 1993:

Safety:

- dermal protection,
- condition and type of work clothes,
- hygiene (water and soap), and
- equipment conditions (leaks).

Effectiveness:

- correct identification of pests and weeds,
- identification of a suitable pesticide for a given problem, and
- dosage amounts.

Table 3.1. Overview of Timeline, Mexico Intervention Programme

1993	1994	1995	1996
Implementation planned for 1993 was held back for a market segmentation on the basis of KAP data. The 1993 comprehensive action plan was reduced to a few trial plots that served to train the field staff rather than communicate with farmers. In winter 1993–94, the project agronomists made some extension in crops other than maize (mainly tomato).	Actual start. Post-test carried out after main season in order to check reach of media and adapt action plan. In winter 1994–95, agronomists were sent to another state. No intervention.	Continuation with the programme on the basis of an adjusted action plan. Programme was expected to end after two seasons of intervention. A KAP study at the end of the season was later overruled due to decision to extend intervention to a third season. Post-test at end of season.	One field agronomist was assigned to head the programme, thus total staff reduced. Endline KAP study carried out at the end of the season.

A Segmented Audience

The main target audience—farmers—was segmented according to basic marketing principles. The dimensions used for this purpose were resource level (skills, education, and economic resources) and resistance to change. Processing of the market research data produced the following profile:

- A small group (6% of the surveyed population) had a high resource level and low resistance to change or, more precisely, a propensity to change. This group could be called *leaders*.
- A second group, called *intruders*, exhibited a relatively low resistance to change and had an average resource level. This group readily accepted new technologies once they were introduced by the leaders. It accounted for 25% of the surveyed population.
- The farmers with few resources could be grouped into two clusters, according to their resistance to change. Those with average resistance to change accounted for 37% of the population and were called *followers*. Those with over-average or high resistance to change accounted for 26% of the population and were called *crusaders*.

Table 3.2. Summary of Mexico Intervention Programme

Component	1993	1994	1995	1996
Human resources				
Field coordinator	1	1	1	1
Agronomists male	4	4	4	3
Agronomist female	1	1	1	1
Equipment for each employee				
Pick-up truck	1	1	1	1
Knapsack	1	2	2	2
Barrel for application water	1	1	1	1
Flip chart	1	1	1	1
Set of safety gear including rubber boots, apron, gloves	1	1	1	1
Slide projector		1	1	1
Loudspeaker mounted on pick-up truck			1	1
Actions for government technicians				
Training courses		X		
Actions for distributors				
Individual visits	X	X	X	X
Group meeting, training		X	X	
Use sales point for mass communication (posters, leaflets)		X	X	X
Involved in distribution of safety gear		X		
Novartis sells safety gear at cost		X		
Ejido commissioners				
Present programme to gain support	X	X	X	X
Support in organizing meetings		X	X	X
Posters in *ejidal* houses		X	X	X
Wives and mothers				
Workshops on safe use		X	X	X
Doctors				
Provision of first aid kit		X	X	X
Course on treatment of intoxications with pesticides		X	X	X
Distribution of manual		X	X	X
Monitoring system of acute cases of pesticide intoxication		X	X	X
Actions for farmers				
Individualized recommendations			X	X
Meetings, workshops		X	X	X
Field days		X	X	X
Demo plots	32	28	24	22
Radio project activity news		X		
Radio spots			X	X
Farmer of the week programme			X	X
Cartoons		X		
Posters		X	X	X
Billboards and signs		X	X	X

Table 3.2. Continued

Component	1993	1994	1995	1996
Actions for children				
Playbook		X	X	X
Actions for agricultural students				
Courses				X

The remaining 6% were relatively large-scale farmers, with a high resistance to change. Due to the small size of this group, they were not given any specific attention.

Intruders controlled 22% of the total maize acreage. They were better educated than the crusaders (secondary level or better), had more economic resources (measured as ownership of television sets, tractors, and refrigerators), used certified seed and seed treatment, and exhibited a positive attitude towards safety gear. They preferred herbicides to hand-weeding. They also had the highest rate of reported intoxications with plant protection products.

Crusaders controlled 21% of the maize acreage. They exhibited a macho attitude; had a primary school education, fewer economic resources, and low agricultural productivity; and were sensitive to the cost of agricultural inputs. On matters of pest control, they were oriented towards pesticides.

Followers controlled 33% of the acreage. They were close to crusaders but, like intruders, had a more positive attitude towards safety. Their farming practices may be called traditional. They had positive attitudes towards safety gear and a somewhat more rational concept of health risks due to pesticides (for example, they did not believe that pesticide operators become immune to the effects of exposure). Their main information source was their own family.

The KAP of these groups regarding the safety and effectiveness of pesticides show some notable differences. (See Table 3.3.)

For practical purposes of the communication campaign, project management only distinguished between two segments—leaders/intruders/followers and crusaders. The appeal of messages directed to crusaders should be powerful, it was decided, and should incite conflict and reflection. The messages directed at leaders, intruders, and followers should be more technical, using rational argumentation.

At the beginning of the programme, there was little information about the appropriateness of the available media for the different segments. Therefore, rather than communicating messages for the different segments through different media channels, communication elements were designed to appeal and attract preferably one or the other segment. (See Table 3.4.) The main media and

Table 3.3. Mexico: KAP of Segmented Audience

Audience	On safety: Knowledge	On safety: Attitudes	On safety: Practice
Crusaders	Low: do not know that dermal exposure may lead to intoxication.	Low or negative: think safety gear are for women; accept the risk of being intoxicated; think that operators become immune; think that pesticides are a blessing.	Low: do not use any safety gear.
Followers	Low	Low	Low
Intruders	Low: do not know about dermal exposure.	High: think that safety gear is for everybody.	Medium: intend to protect themselves.

Audience	On effectiveness: Knowledge	On effectiveness: Attitudes	On effectiveness: Practice
Crusaders	Low: low awareness of pests, no awareness of beneficials.	Positive: need to improve income.	Low: want to kill.
Followers	Low	Low	Very low
Intruders	Low to medium: low awareness of pests and beneficial insects.	Medium: pesticides are not conceived as a blessing; technical advice is welcome.	Medium low: use certified seed and tractors.

slogans, with the exception of slides used at the meetings, were pretested in 1994 using focus group discussions and adjusted accordingly. The slides used at the courses were continously adapted in response to the perceived reactions of the farmers and the gradual shifts in the key messages.

A post-intervention test carried out at the end of 1994 gave some indication as to which segments preferred which media. Project managers detected, for example, that cartoons, which were thought appealing to crusaders and followers, were also read by intruders. The meetings, initially designed to convey knowledge to intruders, were equally well attended by all segments. This was one reason the managers gradually included more motivational elements in the meetings or courses.

Table 3.4. Mexico: Messages Tailored for Segmented Audience

Factor	Leaders, intruders, and followers	Crusaders
Driving force	Fear of health damage Pride of knowledge and skills	Self image (macho) Pride of yield results
Needs	Safety, second-term effectiveness	Cost > effectiveness > safety
Suggested tone of voice	Low key, convince, through influencers	"Bulldozing", direct approach
Suggested media	Fathers of young farmers, TV, radio, technical meetings	Fellow farmer, own experience, wife

Target Groups

Five target groups were identified for the intervention programme:

- farmers—that is, *ejidatarios* growing more than 3 hectares of maize (but no *pequeños propietarios*);
- women—that is, mothers and wives of farmers in the first target group;
- agricultural technicians from both government agencies and distributors;
- distributors and pesticide retailers; and
- medical doctors in both the public and the private sector.

As noted, Villaflores was home to 3,300 farm families, with 5,500 adult male family members engaged in agricultural activities. This was the main target group.

The wives and mothers of the target farmers were thought to represent an access route to pesticide users, to influence their husbands and be respected by them, to be sensitive to rational arguments (that is, not to be caught up in a macho image), to care for health more than men do, to wash the contaminated working clothes of their husbands and sons, and to want to protect their children from pesticide intoxications. Any actions designed for women contained elements directed to children, who were often brought along.

The target group of "technicians" contained some 50 agronomists of the federal and state agricultural ministries and of the federal credit schemes offered in the project area. They did not meet desirable standards of knowledge about safe and effective use of pesticides. They were included as a target group in order to raise their professional level, bring them in line with the communications directed to farmers, and gain their commitment to the project objectives. Some of the technicians might also use pesticides personally and therefore should be motivated to emphasize safety in these tasks. This target group was dropped towards the end of 1994, because their already low presence in the field declined to practically zero when the direct support system (Procampo) was introduced and technicians were used for the administrative tasks relating to that scheme. In 1996, however, some training was provided for agricultural students on a regular basis (five courses at University and nine at agricultural colleges).

In 1994, there were some 35 pesticide distributors in the test area, operating 41 sales points. They played an important role in providing farmers with pesticide-related information, but they often knew little about safe use or even effectiveness. Interventions were designed to improve their knowledge and attitudes, in order to prevent the project from being misunderstood as a commercial one and to enhance the role of distributors as an information and equipment source for safe and effective use.

The medical doctors of the area openly admitted their lack of specific knowledge about treatments of pesticide intoxications. It was felt that this desire for knowledge should be met. Moreover, project managers wanted to gain their commitment to monitor pesticide intoxications.

Training Courses

Courses for Farmers

These courses constituted the key element of the Mexican media plan for several reasons. First of all, they allowed the project to convey technical knowledge and to attract farmers' attention for a reasonable time—an hour, for example. All the *ejidos* have an *ejidal* meeting room, where the training could be held conveniently, using slide projectors and flip charts. These rooms are typically located in the centre of villages, thus facilitating attendance by farmers who do not normally own any motorized vehicle.

All the *ejidos* also have at their disposal a stationary loudspeaker system, which is used as *intraejidal* mass media. The loudspeaker is available to the general public for a small fee, and the messages broadcast include invitations to political events as well as commercial offers from local vegetable traders and the like. Later the project managers found out that the loudspeaker system in some *ejidos* was deficient in that it would not cover the whole village, so they started to use mobile loudspeakers of their own.

The meetings were held once a week in each *ejido*, preferably at 5:00 p.m., when it was easiest for farmers to attend. Measuring cups were distributed free of charge during the courses. In some case, project managers also had the opportunity to present courses at meetings held for other purposes. Attendance was rewarded with a diploma at the end of each season. (See Table 3.5 for totals on attendance.)

The initial main objective of the courses for farmers was to increase knowledge on safety and effectiveness. The courses were designed to improve knowledge, which as noted earlier was low for crusaders and followers, and low to medium for intruders. The managers thought that the degree of motivation to change achieved through courses would depend mainly on the ability of the moderator, and therefore the project trained the field technicians in didactics. They also thought that, due to the rational nature, courses would attract and be effective mainly for intruders (and leaders). Improving the knowledge of intruders would also have an effect on followers and crusaders, since fellow farmers are the

Table 3.5. Mexico: Training Courses for Farmers—Number and Attendance

Measure	Average, 1994–96
Number of courses per season	257
Attendance per season	6,183
Average attendance per course	24
Number of different persons attending	3,000*
Average courses per participant	2.3*

*Evaluated in 1994 only.

most important information source for these segments of the audience. It was not appropriate or possible to invite only intruders to the courses; other segments would attend as well.

After the first season (1994), project managers noted that courses were attended by all segments and that knowledge and awareness levels had been raised considerably. The focus of the courses was therefore shifted to motivation (see Table 3.6), done through the inclusion of more practical elements; more courses in the demo plots (the field days), where the benefits were visible; and the addition of intoxication testimonies by farmers (in 1995) and medical doctors (in 1996).

Courses for Women

The goals here were to motivate women to influence the pesticide use practices of their husbands and sons and to provide the necessary knowledge for them to do so. Women's KAP regarding activities during which they might come into contact with pesticides (storage, washing clothes, treatment of intoxicated husband) were to be improved. At these meetings, a children's playbook was also distributed. Slides and flip charts were used to present information on safety, prevention, handling of pesticides, personal protection equipment, and hygiene habits.

The courses for women were given by the part-time female agronomist employed by the project. *Ejido* leaders provided the meeting rooms and took care of the invitations, along with the presidents of local women's associations. The courses were also announced in the radio programmes and spots, via loudspeaker, and at events of DIF extension workers. (DIF is a government organization with social objectives, which cares for family development.)

Courses for Government Technicians

These courses covered weed control, integrated pest management (IPM), knapsack sprayer calibration and pesticide dosages, knapsack sprayer maintenance and repair, and first aid and treatment of accidental intoxications. They were given by the project field coordinator. As mentioned earlier, they were abandoned in 1994 due to the unavailability of the technicians.

Table 3.6. Mexico: Training Courses for Farmers—Topics and Materials

1994	1995	1996
Knapsack sprayer maintenance and repair	Knapsack sprayer maintenance and repair, safety key issues	Dropped
Knapsack sprayer calibration and product dosage	Calibration and product dosage	Dropped
Safety equipment and hygiene habits	Safety equipment and hygiene habits, intoxication testimonies, good and bad practices of community members	Safety, doctors' testimonials, good and bad practices of community members
First aid and treatment of intoxications	Dropped	
Environmental protection related to pesticides	Dropped	
Weed identification and control (in maize)	Herbicides: effectiveness key issues, application technique, rating, demo plot results	Weed control, including calibration, dosage, demo plot results
Insect pest identification and control (in maize)	Insecticides: effectiveness key issues, application technique, rating, demo plot results	Insect pest control, including calibration, dosage, demo plot results

Courses for Distributors and Technicians

These courses dealt with herbicide use, IPM, knapsack sprayer calibration and pesticide dosages, knapsack sprayer maintenance and repair, and first aid and treatment of accidental intoxications. Materials posted at the sales points include posters showing safety measures. The male field technicians were in charge of the courses to distributors.

Courses for Doctors

The themes of these courses were first aid and treatment of intoxications, knowledge about types of pesticides, and the likely symptoms of intoxication. Treatment kits, manuals, and posters were supplied free of charge. The courses were given by a medical doctor of the industry organization AMIFAC (Mexican Association of the Phytosanitation Industry, formerly AMIPFAC), which also provided materials. AMIPFAC guaranteed replacement of antidotes on request and without cost to doctors. A form to monitor treated pesticide intoxications was also distributed to doctors.

Other Components of Intervention

Demo Plots

On the demo plots, project technicians implemented Novartis's crop protection programme on one part of a farmer's field; this consisted of applications of soil insecticides, preemergence herbicides, foliar insecticides, and postemergence herbicides where required. In 1994, the project agronomists carried out the applications of pesticides. In 1995 and 1996, the farmers applied the products themselves, under the supervision of the agronomists. This was done in order to increase the motivational and didactical value of the demo plots. Usually, one demo plot was installed in each *ejido*, and marked with a sign. The demo plots covered 0.10–0.25 hectares.

Yields and cost of inputs were evaluated at the end of each season. Typically, yields in the demo plots were about 20% higher than in the remainder of the fields. Input costs were also higher, but they were more than compensated for by the increase in yields. These results were communicated to the farmers at the field days, at meetings, and during individual contacts.

Demo plots were included in the communication campaign in an effort to reach farmers who believe only what they see. The testimony of the farmers who participated in these would be that of a "fellow farmer" to crusaders and followers, who prefer this information source. Demo plots were expected to create curiosity and interest. This was confirmed by the technicians, who were often contacted by farmers for advice about safe and effective use of pesticides, and about agronomic problems in general. Thus it seemed that demo plots would be the main channel to reach followers and crusaders, who are more likely to accept changes yielding an economic benefit.

The primary objective of demo plots was to demonstrate effective use of plant protection products. But they also served to demonstrate safe use and as locations for field days. Project managers soon noted that the major value of demo plots was that of creating trust by the farmer as he observed that the project had something tangible to offer (economic benefits). This facilitated the communication of safety messages.

Field Days

Field days are didactically more suitable than training courses for conveying both knowledge and motivation (attitude change). They allowed inclusion of some practical training elements that could hardly be done at the meetings in the *ejidal* meeting rooms. (See Table 3.7.) But they were also logistically more difficult to organize. In each of the three years of the project, between 20 and 40 field days were held, with a total attendance of some 2,000 farmers.

Table 3.7. Mexico: Topics of Field Days

Stage	1994	1995	1996
At sowing		Products used at start of season. Farmers apply products themselves under instruction.	Plant protection and safety. Products used at start of season.
Mid-season	Insect pest identification and control, weed identification and control, safety	Demonstration of exposure at application. Safety. Demonstration of results of herbicides.	Dropped
Late season		Show benefits: Summary of demo and control plot techniques, conclusion of programme with most active farmers, distribution of diploma for attendance.	Show benefits of implemented management techniques.

"Farmer of the Week" Programme

With the objective of motivating farmers to improve safety, the field coordinator randomly picked farmers found spraying. The farmer's practices were evaluated, discussed, and awarded according to a detailed score table.

Prizes for good practices included nozzles, watches, rubber boots and gloves, knapsack gaskets, and knapsack sprayers. Applicators showing relatively high safety standards were honoured in the radio programme as "the farmer of the week".

This programme was designed for followers and crusaders—groups that respond to an "ordering" tone of voice. They were informed about the "farmer of the week" programme through radio, family members, and other farmers.

Cartoon Book

Although Mexico cannot be called a nation of readers, cartoon books are a traditional type of literature (called *revistas* or *historietas*) and are commonly read by the relatively poor. The characters are human and the themes are like soap operas.

This communication channel seemed to be particularly suitable for reaching individuals not particularly interested in safety and effectiveness: it packed our messages into a context of love and ambition. The objective of the cartoon booklet was to motivate followers and crusaders in an emotional way, enhance their ambitions to be good farmers, break the image that safety gear is unmanly, associate safety with a macho image and with effectiveness, and improve knowledge with easily understandable technical information. Originally, three different comic books were planned, with different episodes of the same story. Later these were combined into a single booklet.

The booklets were distributed at farmers' and women's meetings. Their characters appeared on some of the posters as well.

Radio

In 1994, the local radio station was used to broadcast "project activity news". This consisted of a weekly 15-minute programme, aired twice, that described the activities of the field technicians during the previous and following weeks, upcoming courses, campaign messages, intoxication testimonies, technical support, the "farmer of the week", and music. The objectives were to inform the whole population about the project, communicate the main messages, support other actions, motivate through continuous presence, and improve technical knowledge.

A survey conducted to estimate the reach of radio found that 78% of the farmers listened to the only station in the project area. (This station could not be heard in the control area of Cintalapa until 1995, when it increased its power and the programme was therefore stopped.)

Billboards

The billboards showed the following message:

Care for your family
Care for your health
Care for your crop

They also included the logo of the Novartis Foundation for Sustainable Development. Ten billboards were posted at crossroads and access roads to major *ejidos*.

Playbook

The objectives of giving children a playbook were to teach future farmers basic safety rules, communicate that pesticide safety is a family matter, influence farmers who do not care for themselves but do care about their children (crusaders), and motivate women to attend meetings. The playbook included mazes, pictures to be painted out, pictures identifying safety errors, and so on. It was distributed at the courses for women.

Posters

Three posters were produced. One said:

We love you strong,
we love you healthy,
we love you being ours,
we love you happy,
we love you very much

These words were accompanied by pictures of five local people, including women, children, and elderly individuals, thus representing the family members of the average farmer. Pictures of recommended safety practices were positioned along one side of the poster. Thus, the poster conveyed a life-affirming message and associated it with the family and the safe use of pesticides.

The other two posters showed characters from the cartoon book and the key messages relating to safety and effectiveness.

Monitoring Cases of Intoxication

The project hoped to collect information on cases of pesticide intoxication that were treated by doctors. A form to monitor such treatment was distributed to all doctors. Completed forms were periodically collected by the project staff. The seasonal total of cases reported did not exceed 14 in any one year. Of these, 8–10 were normally suicide attempts. The number of occupational intoxication cases reported by doctors was thus very low and did not allow any conclusions to be drawn about whether this problem was increasing or declining.

Programme Adjustments

Year One (1994)

In January 1994, the world became aware of Chiapas as the Zapatista National Liberation Army occupied a handful of villages in the Chiapas highlands. The war did not directly affect the intervention area, which is located about a three-hour drive from San Cristóbal, the largest and closest of the affected cities. The fires ceased two weeks later, and project management decided to continue the programme as planned. Although the war caused political unrest in Chiapas, mainly during 1994 (an election year), the practical impact on the programme turned out to be minor.

Year Two (1995)

A post-test of a sample of 1,000 farmers (500 each in the test and control areas) was conducted after the first season of intervention (1994), and the results were extrapolated to the entire target population. (See Table 3.8.) The survey found that 72% of the recalled messages stemmed from the courses and field days, while the demo plots contributed 13%, the radio programme 5%, the cartoon 4%, and the posters 6%.

Table 3.8. Mexico: Results of Communication Campaign After One Year

Action	Contacts	Reach	Contacts per reached grower
Courses and field days	9,800	2,800	3.5
Demo plots	n.a.	3,500	n.a.
Radio programme	4,250	1,150	3.7
Cartoon	2,140	880 (read book)	2.4
Posters	n.a.	3,200	n.a.
Overall	> 22,890	4,350	> 5.3
Average per male adult target population	> 4.16	79%	
Average per target household	> 6.94	> 79%	

Action	Reach	Average recall (number of themes)	Total recall (number of themes)
Courses and field days	2,800	4.80	13,440
Demo plots	3,500	.70	2,450
Radio programme	1,150	.78	890
Cartoon book	880	.85	750
Posters	3,200	.35	1,120
Overall	4,350	4.28	18,650
Average per male adult target population	79%		3.39
Average per target household	> 79%		5.64

The study showed that the reach of the programme in its first year was high. Four out of five members (80%) of the main target group had heard about the programme and recalled an average of 4.3 themes communicated by various media. For each recalled recommendation, 65% of the recallers claimed to put it in practice. Due to the dilution over many messages, however, this translated into relatively low (claimed) improvements in practices. Project managers concluded that in 1995 the motivation to put what is learned into practice had to be emphasized, that attendance at the meetings had to be increased, and that the project needed to focus on fewer key messages.

At the end of 1994, health models were developed using the KAP data of Mexico and Zimbabwe. They showed that washing and avoiding knapsack sprayer leaks would yield large and relatively certain effects in slowing health damage.

This had been expected, of course. But the magnitude of these effects was not anticipated. Leaking sprayers and negligence regarding hygiene were found to triple risks, while many other precautions yielded rather ambiguous results. Even if the latter were due to problems of methodology, the findings reinforced the decision to focus on the most important precautions and on a few strong, key messages, and to refrain from recommending practices that may be useful but are unlikely to be adopted by a large share of farmers. (See Table 3.9.)

Table 3.9. Mexico: Evolution of Key Messages, 1994–96

	1994	1995	1996
On safety	Dermal protection Condition and type of work clothes Hygiene Equipment conditions	Bathing after spraying Knapsack sprayer maintenance Footwear	Bathing after spraying Knapsack sprayer maintenance Footwear Use of measuring cups
On effectiveness	Identification of pests and weeds Specific product for specific problem Correct rate	Specific product for specific problem Application timing Correct rate	Specific product for specific problem Application timing Correct rate

Several corrective actions were taken at the beginning of the 1995 growing season. First, in order to improve motivation for change, project managers introduced:

- individualized communications,
- more testimonies on intoxication at farmers' meetings,
- new posters,
- more "ready to use" recommendations instead of more general and abstract information, and
- pesticide applications in demo plots done by farmers themselves (instead of by project technicians).

Second, in order to increase the reach of the programme, there was:

- better promotion of meetings, using mobile loudspeakers;
- more meetings for growers;
- more field days;
- more motivation and less information;
- new posters and better placement in different places; and
- more playbooks.

Third, in order to improve the audience's recall of main points, management instituted:

- a strict focus on key messages,
- more practical elements in courses and field days, and
- fewer topics in courses.

Last, in order to save resources used in these activities, we:

- reduced demo plot size,
- reduced actions for retailers, and
- held no further activities for technicians.

In December 1994, the peso suffered a drastic devaluation. This caused an inflationary shock (30–60% increase in a few months) on the prices of agricultural inputs that are imported or otherwise linked to international prices (such as pesticides and fertilizer). The price of consumer goods also rose. Maize prices did not follow the same pattern. By early 1995 they had decreased 20% nominally and around 30% in real terms since early 1994, and were below international levels.

These events led the average maize grower to reduce input levels drastically, while growers with inferior production techniques abandoned maize entirely or, if they relied on it for subsistence, stopped using external inputs. In 1995, therefore, pesticide use dropped in both the test and the control area. Moreover, farmers started to shift back towards cheap and toxic products such as methyl parathion and paraquat. Under such adverse economic conditions, an increased resistance to change could be expected, because the farmers' capacity to absorb the (subjective) risk involved in innovation was reduced.

Year Three (1996)

Following the 1995 adjustments to the programme, a second post-test was conducted. It showed that the measures taken at the beginning of 1995 indeed improved reach and recall. Only minor changes were made to the programme, therefore, basically reinforcing the changes made in 1995.

By contrast to 1995, maize prices climbed in 1996 to 1,600–1,800 pesos (more than US$200 per ton). MASECA, the major maize mill, offered a credit scheme in the area. Unlike in other years, governmental direct support and credit schemes arrived on time. All these factors contributed to a maize boom. Agriculture seemed to be highly profitable—as illustrated, for example, by the fact that the number of private agricultural consultant firms in the test area rose fourfold, from two to eight. But the boom came to an abrupt end at harvest time, when the maize price actually paid settled at 1,270 pesos per ton. In protest, farmers in the test area blocked roads for several weeks, isolating Villaflores from trade in the rest of the state.

Results of Intervention

As noted earlier, the programme was evaluated by means of a qualitative study (focus group sessions) and quantitative studies, consisting of the KAP and observation surveys. The results of these evaluations are presented in this section.

Qualitative Evaluation

The qualitative evaluation of the results of the intervention programme consisted of focus group sessions with farmers—eight sessions in Villaflores and two in Cintalapa. As with the initial focus groups, these sessions typically had 8–10 farmers in each one, and were guided by a motivational psychologist. They were held 9–14 September 1996. The participants were selected with due consideration to balance in terms of their ages, symptoms of pesticide poisoning within the last two years, and (in the test area) attendance at the programme's training courses.

The farmers' willingness to participate in the focus groups was always high, and it increased with time. The discussions about agriculture were generally not distracted by any other topics with the exception of the midterm evaluation, when farmers in Villaflores worried about the damage caused by an earthquake that destroyed about 10% of the buildings in October 1995.

Although a military presence increased suddenly and visibly in both areas after January 1994, the Zapatista uprising was geographically distant and only of marginal interest to the farmers of Villaflores and Cintalapa. Partly as a result of the Zapatista movement in the Chiapas highlands, political instability did increase in other regions of the state, but apparently it was not very important to the farmers in the project areas. Local political events would occasionally affect the work of the project team, however, by distracting the farmers' attention and reducing attendance at meetings.

A certain "urbanization" became increasingly apparent over the intervention period in Villaflores. The mass media expanded its reach and increasingly influenced perceptions about pesticides. This communication was reinforced by local opinion leaders, such as teachers and young professionals.

Impact on Knowledge

The farmers who had attended courses showed an increased level of knowledge of plant protection products and of the rules of safe and effective use—they expressed no doubts about the way pesticides should be used, they correctly differentiated between classes of herbicides, and they used universal terminology as well as local names to describe pest problems (although they still preferred the latter). In addition, the number of participants with considerable knowledge of these topics was larger than in 1992.

Even the farmers in the test area who did not attend courses knew more about pesticides than they did in 1992. Apparently information had passed from participant to nonparticipant. The knowledge base of these farmers was weaker, however, and the use of local terms even more pronounced.

Impact on Attitudes and Practices

The changes apparent in some farmers went far beyond the acquisition of knowledge about safety precautions, effectiveness, and recommended practices. (See Table 3.10.) In particular, the following changes were noted:

- changes in the values that direct individual behaviour and conduct, with the affirmation of life predominant;
- increased self-esteem, which reduced the macho attitudes and behaviour detected in 1992;
- a move away from a passive, receptive attitude and a feeling of dependence on outside factors, such as the government or the effectiveness of a particular pesticide; and
- a new perception of reality, with farmers being accountable for crop productivity.

Table 3.10. Mexico: Profile of Farmers Who Changed the Most

1992	1996
Farmers acted according to and depending on traditional practices, based on experience gained by trial and error.	Farmers use the information provided by trained technicians.
Pesticide selection based on limited experience; untrained distributors influenced decisions of farmers who lack self-confidence.	Pesticide selection based on knowledge farmers acquired or on recommendations of trained agronomists.
Macho attitude towards use of plant protection products.	Awareness of vulnerability; willingness to protect themselves.
When crop failed, outside forces and persons held responsible.	Awareness of ability to influence productivity.

Thus one of the most important differences between the test and control areas at the end of the study concerns the attitude towards the perceived obstacles to successful farming. While farmers in Cintalapa tended to blame the government for economical constraints and felt incapable in the face of pest problems, farmers in Villaflores who attended the project courses not only described the problems in more detail, they could also identify potential solutions. They showed increased confidence regarding pesticides and considered pests and weeds manageable, resulting in a reduction of the importance attributed to the pesticide price or cost. Test area farmers perceived pesticides as the most effective means of pest control, while control area farmers perceived them as the only means. The soil conservation technique of not burning stubble before sowing was intensively discussed among farmers who had been at the courses, while farmers in Cintalapa were hardly able to look beyond the practical obstacles to adoption of that practice.

Awareness of the health risks involved in pesticide use was high among Villaflores farmers who had experienced related health problems. These individuals think that the health risk is a result of improper use rather than an inherent pesticide attribute. Yet not all of these farmers would follow the safety rules they knew.

Perhaps the most dramatic change between the start and the end of the study was the test area farmers' attitudes towards their own body and health. At the baseline, the nonuse of protective items was accompanied by what could be called a "macho" attitude that prevented farmers from admitting their physical vulnerability. They tended to attribute early, minor intoxication symptoms to everything but pesticides, given that they thought only the weak can suffer intoxication. At the end of the study, farmers accepted that they are at risk, were able to recognize early intoxication symptoms as such, and, as a consequence of this, tried to protect themselves despite economic constraints and mostly by simple means, such as wearing boots and long-sleeved shirts, putting a wet handkerchief on the face, or taking the wind direction into account.

At the endline evaluation, test area farmers also showed a more rational attitude towards pesticides, as expressed in a number of practices, including

- rotation of chemical classes of insecticides,
- checking the expiration date when buying pesticides,
- storage of the products outside the house,
- avoidance of dermal contact with pesticides,
- intent to improve pesticide effectiveness by timing the application,
- use of the approximate recommended dose,
- not allowing minors to spray,
- washing spraying equipment after use in order to increase its durability,
- changing fittings, and
- greasing the equipment.

Thus farmers became more careful and thoughtful with regard to their own health and that of their children and their crops, as well as with their equipment. Few improvements were detected with regard to the disposal of empty pesticide containers, however.

Overall Impact

Farmers who attended the training courses often referred in the focus groups to the yields and health benefits achieved through the programme. The health impacts were mainly brought up by those who had suffered from pesticide poisoning.

The satisfaction level was markedly higher among the farmers who had followed the programme's recommendations for several seasons and therefore were better able to appreciate the benefits. Compliance with safety rules was also found to contribute to satisfaction level.

The major reasons given for not adopting safety practices were lack of economic resources, the inadequacy of some protective gear under prevailing climatic conditions, and the simple fact the change takes time—that is, greater changes would have been observed if the programme had continued. The researcher also noted that the language used by the staff was occasionally too technical, which hindered the farmer's understanding.

The farmers in Villaflores who did not attend any training courses also changed their attitudes towards those who adopted safety precautions: they no longer criticized them.

In Cintalapa, the control area, the attitudes towards pesticides and the practices were basically the same in 1996 as in 1992.

Quantitative Evaluation

Methodology

During the four years between the baseline and endline surveys, behaviour changed for many more or less obvious reasons. As mentioned earlier, it was assumed that these changes overlap the project-induced adoption process in the test area, but can be detected in the control area. Therefore, the programme effect is assumed to be the difference between the changes in the test and control areas, or the "net change". In addition to the baseline and endline surveys, a follow-up survey was conducted one year after cessation of the intervention programme, with the hope of assessing the persistence of any changes detected earlier. Thus for any variable assessed all three times, there is one net change between the baseline and the endline (Net) and another change between the endline and the follow-up (Net+).

Net, then, quantifies the project-induced incremental adoption—that is, the proportion of farmers in the test area who have adopted a certain piece of knowledge, attitude, or practice as a result of the project. Net+ quantifies the persistence of adoption. A positive Net+ indicates that adoption in the test area increased after conclusion of the project (in comparison with the control area). This could occur where communication between farmers continues to spread the knowledge, attitudes, or practices in question. It could also be the result of farmers experimenting with new practices on a small scale before adopting them completely. A Net+ value near 0 indicates that adoption stayed approximately at the level achieved when the intervention stopped, and a negative Net+ value indicates that adoption decreased after cessation of the project (always in comparison with the respective control area development).

One disadvantage of this method is that it depends crucially on the assumption that the outside agents that affected the control area were active to the same extent in the test area. This assumption was probably violated in the Mexican project, however. Therefore the absolute changes in the test area, without consideration of the control area, should and will also be analysed.

Media Reach

Nearly two in five (37%) of the test area growers spontaneously mentioned having attended meetings when asked whether and from what source they had got pesticide-related information. Further analysis showed that most of these growers referred to courses organized by Novartis. On direct questions, farmers confirmed having seen pesticide-related posters (43%), cartoon books (42%), and demo plots (44%); 51% recalled having been to meetings or courses of the Novartis Foundation. About one third of these growers attended five or more events and therefore were likely to have been exposed to most of the topics presented.

Adoption of New Attitudes and Behaviour

Tables 3.11 to 3.15 provide overviews of the values at different times of a number of variables assessed in the KAP and observation surveys. It is worth repeating that Net+ indicates the change between the endline and follow-up surveys, not between the original baseline survey and the follow-up.

Summary of Major Findings

Continued Changes Due to the Project

A detailed analysis of the study's findings in Mexico yielded a group of KAP items that are most certain to have been persistently adopted due to the project. They include:

- the practice of washing hands before eating in the field during a break in the spraying job,
- the use of pesticide measuring tools other than the formerly predominant chili can,
- the practice of cleaning the knapsack sprayer after use,
- the practice of putting on rubber boots right at the beginning of a day of pesticide spraying,
- the use of a long-sleeved shirt while spraying,
- the practice of not throwing empty pesticide containers away in the field,
- the practice of keeping pesticides in their original containers,
- the knowledge that pesticide toxicity is indicated by a coloured stripe on the container, and
- storage of pesticides out of children's reach.

Table 3.11. Mexico: KAP Survey Results on Knowledge (percentage)

Knowledge	Test area			Control area			Test area changes		Net changes	
	B	E	F	B	E	F	E-B	F-E	Net	Net+
Pesticides not harmless[a]	90	94	100	95	93	100	4	6	6	-1
Toxicity indicated by stripe[b]	12	19	27	10	6	12	7	8	11	2
Boots[c]	36	37	43	18	37	24	1	6	-18	19
Gloves at mixing[c]	33	44	38	36	51	31	11	-6	-4	14
Apron[c]	8	17	23	29	14	36	9	6	24	-16
Beneficials[d]	9	21	25	19	25	39	12	4	6	-10

B = baseline E = endline F = follow-up

[a] All farmers but those who affirmed that pesticides are not harmful to humans (or similar) when asked who faces the largest pesticide-related health risk.

[b] Farmer mentions a coloured stripe when asked how toxicity is indicated on the label.

[c] Farmer mentions specified protection when asked about ideal protective measures to take when spraying the most toxic pesticide he has ever used.

[d] Farmer affirms that some insects are beneficial when asked whether all are harmful or some are beneficial.

Table 3.12. Mexico: KAP Survey Results on Attitudes (percentage)

Attitude	Test area			Control area			Test area changes		Net changes	
	B	E	F	B	E	F	E-B	F-E	Net	Net+
Pests not #1 problem[a]	19	32	57	38	33	66	13	25	18	-8
Effectiveness first[b]	29	49	32	44	47	29	20	-17	17	1
Safety first[b]	63	28	26	44	29	25	-35	-2	-20	2
Safety not third[c]	89	57	51	76	57	47	-32	-6	-13	4

B = baseline E = endline F = follow-up

[a] When asked about agricultural problems in general, farmer did not mention pests or weeds first.

[b] Ranking when farmers were asked to rank "effectiveness", "no harm to humans", and "cost" by importance.

[c] When asked to rank "effectiveness", "no harm to humans", and "cost", farmer ranked "no harm to humans" first or second.

Table 3.13. Mexico: KAP Survey Results on Reported Practices (percentage)

Practice	Test area			Control area			Test area changes		Net changes	
	B	E	F	B	E	F	E-B	F-E	Net	Net+
Rubber boots[a]	43	76	62	14	43	33	33	-14	4	-4
Gloves[a]	16	12	12	11	16	17	-4	0	-9	-1
Apron[a]	10	14	18	21	18	32	4	4	7	-10
Wash hands before eating[b]	67	93	97	82	86	88	26	4	22	2
Uses dosifier[c]	0	34	36	0	25	21	34	2	9	6
Does not use chili can[d]	30	59	58	40	39	36	29	-1	30	2
Cleans sprayer after use[e]	70	85	77	91	93	92	15	-8	13	-7
No sprayer leaks[f]	88	97	92	92	96	91	9	-5	5	0
Pesticide out of children's reach[g]	18	25	26	23	29	24	7	1	1	6

B = baseline E = endline F = follow-up

[a] Farmer spontaneously mentions specified protection when asked protection measures taken on latest spraying occasion.

[b] Farmer spontaneously mentions handwashing when asked whether they took precautions before eating, drinking, or smoking during latest spraying, expressed as proportion of all respondents who took breaks for eating, drinking, or smoking.

[c] Farmer reports using a measuring tool when asked how he measured pesticide quantity.

[d] Farmer mentions anything but a chili can when asked how he measured pesticide quantity.

[e] Farmer affirms cleaning sprayer after each use.

[f] Farmer does not mention leaks when asked condition of his knapsack sprayer.

[g] Farmer answers "out of children's reach" when asked where pesticides stored between purchase and use. (Note that other answers refer to the storage place, which may also be out of children's reach; thus the variable captures partly the awareness that pesticides should be kept away from children, and partly the practice.)

Table 3.14. Mexico: KAP Survey Results on Health (percentage)

Health indicator	Test area			Control area			Test area changes		Net changes	
	B	E	F	B	E	F	E-B	F-E	Net	Net+
Never been ill due to pesticides	75	63		88	75		-12		1	
No health effect at latest spraying	97	93		98	95		-4		-1	

B = baseline E = endline F = follow-up

Table 3.15. Mexico: Observation Survey Results on Practices (percentage)

Practice	Test area			Control area			Test area changes		Net changes	
	B	E	F	B	E	F	E-B	F-E	Net	Net+
Rubber boots on way to field[a]	12	48	46	16	21	17	36	-2	31	2
Rubber boots while mixing[a]	14	61	49	16	30	27	47	-12	33	-9
Rubber boots while spraying[a]	20	68	55	18	41	37	48	-13	25	-9
Not spraying barefoot[a]	86	92	95	98	83	88	6	3	21	-2
Gloves while mixing[a]	0	2	1	0	6	2	2	-1	-4	3
Apron while spraying[a]	15	13	6	0	13	14	-2	-7	-15	-8
Face mask while spraying[a]	22	38	20	15	22	16	16	-18	9	-12
Long shirt while spraying[a]	70	95	92	76	92	88	25	-3	9	1
Clean-looking clothes[a]	14	44	49	40	41	56	30	5	29	-10
Did not use chili can[b]	6	53	52	18	45	37	47	-1	20	7
Used dosifier[b]	1	38	44	2	33	24	37	6	6	15
Washed hands before eating[b,c]	17	71		24	66		54		12	
Hygiene after spraying[b,d]	72	74	68	78	75	66	2	-6	5	3
No skin contact with pesticide noted while mixing	92	91		96	83		-1		12	
Pesticide not repacked[e]	82	87	92	96	98	98	5	5	3	5
Mixing not done by minor[b]	85	80	92	96	93	93	-5	12	-2	12
No minors present[b]	87	78	81	96	91	84	-9	3	-4	10
Cleaned sprayer after use[b]	9	56	71	46	86	81	47	15	7	20
No sprayer leaks noted	78	89		84	92		11		3	
No empty packs in field noted	12	36	41	42	49	49	24	5	17	5

B = baseline E = endline F = follow-up

[a] Observed attire.

[b] Observed practice.

[c] Percentage of those who ate in the field.

[d] Combination of various hygiene measures, such as washing hands or body.

[e] Pesticide was in its original container.

The magnitude of adoption of most of these variables is considerable, being in the range of 20–60% in comparison with the baseline and on the order of 10–30% in comparison with incremental adoption in the control area.

Other items also changed favourably during project implementation in comparison to both the baseline and the control area. The increase continued after implementation, but at a slower rate than in the control area. These items include:

- the frequent washing of work clothes used for spraying,
- the practice of not spraying barefooted,
- knowledge about the existence of beneficial insects,
- knowledge about the importance of aprons,
- knowledge that pesticides are not harmless to humans, and
- the perception that pests are not the number one problem in agriculture.

These items are also very likely to have been adopted persistently. With the exception of the practice of washing the work clothes frequently and not spraying barefooted, incremental adoption of most of these items was rather small, although in the case of the knowledge that pesticides are not harmless, this was due to the already very high baseline level. In the case of knowledge about beneficial insects, the small increase is related to the fact that beneficials were a low-priority topic for the project in comparison with the more urgent key messages.

Desired Changes That Did Not Continue

Another group of behaviours increased during but decreased after the implementation, whether compared with the baseline or the control area. Thus these changes were adopted due to the project, but not persistently. Some of them were well adopted and tend to be abandoned only slowly, and their adoption level is still favourable in comparison with the baseline or the control area. Included in this group was the use of rubber boots while spraying and mixing pesticides, as indicated by several variables in the observation and the KAP surveys. Some farmers wear boots during mixing and spraying but not on the way to the field. Apparently, they carry boots to the field and change into them before starting to spray. It is these growers who were most likely to abandon the practice of wearing boots between the endline study and the follow-up study done a year later.

Some knowledge and practices were completely or partially forgotten or abandoned after having been modestly adopted during the intervention phase. This was the case for the use of face masks and the practice of maintaining the knapsack sprayer in good condition. The stated practice of cleaning the sprayer after each use also falls into this category, but the observed practice changed favourably, as mentioned earlier.

Changes in Undesirable Direction

The variables that changed unfavourably, whether compared with the baseline or the control area, include the observed use of aprons, the reported and the observed use of gloves, and the ranking of safety among the pesticide attributes effectiveness, safety, and cost. The lower ranking of safety was accompanied by an increase in the ranking of effectiveness.

Other Changes

Some of the remaining variables showed an increase after but not during the project implementation. These are difficult to interpret as being the result of the project. This is the case for the avoidance of spills on the skin, which stayed at its initial level in the test area but dropped in the control area, leading to a favourable but ambiguous net change. The knowledge about the protective effect of boots also stayed at its initial level in the test area but increased in the control area, implying an unfavourable but equally ambiguous net change. The involvement of minors in pesticide usage increased initially and dropped later. This development can hardly be attributed to the project.

Health

No health improvement is apparent from these data, whether measured as the incidence of health problems after the latest spraying prior to the interview or as the percentage of farmers who stated that they had ever suffered ill health due to pesticides. To the contrary, the proportion of farmers who report health problems actually increased. As argued in the next section, this is due to cognitive changes rather than an increase in physical problems.

Comparison of Qualitative and Quantitative Results

Generally, the qualitative and quantitative surveys coincide with regard to safety practices. The focus group sessions had indicated that mainly simple gear, adapted to the local climatic conditions, was used. This was confirmed by the quantitative survey, which indicated that practices such as washing hands and equipment and the use of suitable measuring tools and long-sleeved shirts were adopted both more and more persistently than the use of rubber boots, which in turn was adopted more than gloves and aprons were. It would seem that gloves, aprons, and possibly boots are quite wearisome, while washing is refreshing. The two survey results also coincide with regards to pesticide storage and cleaning of the spraying equipment. There are some contradictions between the results, but these mostly involve issues that changed only little.

As to attitudes, the motivational analyst's finding that test area farmers started to consider pests and weeds manageable was paralleled in the reduced perception of pests and weeds as the number one agricultural problem. Similarly, the lower ranking of the safety attribute of a pesticide might be related to the fact that farmers increasingly feel that the health risk is due to improper use rather than being an inherent characteristic of a pesticide. The growing importance attributed to the effectiveness of a pesticide may be related to the improved ability to distinguish between different types of pesticides.

The focus group sessions documented increased awareness of health risks and increased readiness to admit physical vulnerability. This is likely to be the reason that more health problems were reported despite the considerable number of preventive measures that were adopted. Farmers learned how to recognize even minor symptoms and to admit having suffered them, and recalled more incidents of ill health due to pesticides when asked about them. This is underlined by the fact that although the number of reported health problems increased between the baseline and the endline, the average time that farmers rested due to pesticide-related illness was reduced from 0.32 to 0.22 days per season in the test area, while it remained stable at 0.32 days per season in the control area, according to separate surveys carried out to assess the severity of reported health problems. Thus the health problems reported at the endline were probably much less severe, on average, than those reported at the baseline.

Conclusions Regarding Project Hypotheses

As described in Chapter 1, based on several basic facts regarding safety and effectiveness of pesticide use, the Steering Committee established four hypotheses to be tested during the project. These can now be reviewed in the light of the findings in Mexico.

Hypothesis: Through communication and training we will improve practice in:

- skin protection,
- preparation of spray solution,
- washing of body and work clothes,
- spraying and application, and
- maintenance of spraying equipment.

Personal safety is the area where major improvements were detected in Mexico. Persistent changes were found in the use of shirts, boots, and footwear; in the washing of work clothes and hands; in sprayer cleaning; and in avoidance of the chili can as a measuring device. Major attitude changes were detected at focus group sessions. The project therefore had a favourable impact on comparatively simple, cheap safety practices, while more cumbersome practices did not change for long or were used even less.

Hypothesis: Through communication and training we will improve practice in:

- optimization of quantity of plant protection products used and of spray parameters,
- storage of plant protection products, and
- disposal of empty containers.

This area showed some favourable changes, but to a lesser extent than in the personal safety area. The changes that persisted after the intervention programme ended include not repacking the pesticide package, storing pesticides out of children's reach, and improved disposal of empty containers.

Hypothesis: Through communication and training we will improve practice in:

- identification of pests and beneficial insects,
- selection of suitable product,
- determination of correct dosage,
- usage of suitable equipment,
- correct timing of application, and
- proper application techniques.

There were relatively few detectable effects in the area of pest identification and product selection. Knowledge of beneficial insects improved, and the focus group sessions detected more rational attitudes towards pesticides. However, no yield or productivity effects were detected by the KAP studies even though the demonstration plots managed by the project technicians produced superior crop yields.

Hypothesis: Improvement of farmers' economics will facilitate their adoption of messages on safety.

It is difficult to prove this hypothesis with a KAP or observation survey. Evaluation should instead be based on the results of qualitative surveys and the experience of local project staff. Still, it seemed that the demonstration of more-productive crop protection techniques on the plots managed by the project staff sparked farmers' interest and raised their level of trust, which in turn facilitated communication. More rational attitudes towards pesticides affected farmers' perception of both pesticide effectiveness and safety issues. Otherwise, however, there does not seem to be a strong link between these issues, as safer practices were adopted despite a lack of noticeable improvements in effectiveness.

Overall Conclusions

The results presented showed that a number of the promoted practices were adopted and that this adoption was closely related to farmers' changed perceptions of pesticides and of their own bodies and health. Few knowledge variables showed wide adoption, however. Rather, increased knowledge was observed in the form of more-detailed explanations given during the focus group sessions and the certainty with which answers were given. Similarly, the average number of answers given to open questions of the KAP surveys increased over time. It could be said that the quality of knowledge increased rather than its quantity. The quantitative surveys may not have been particularly suitable for eliciting this kind of improvement.

Most of the protective safety practices promoted were quite simple, and some more or less diffuse knowledge about their adequacy existed even at the baseline. What the project achieved was to affirm and partly correct this knowledge. This then helped overcome inadequate perceptions and attitudes, which had been obstacles to change.

Jointly, the qualitative and quantitative evaluations showed that the project achieved a large number of desired changes regarding knowledge, attitudes, and practices. It is highly probable that these changes had a favourable effect on farmers' health. Unfortunately, the health variables assessed did not help capture this effect because they are affected by attitudes as much as by underlying physical realities.

4

India

Overview of the Agricultural Sector

The large majority of Indians depend on agriculture for a living. In 1995 about 57% of total land area was cropland, a ratio unchanged since 1980. The inelasticity of land supply means that increasing output will depend on improved use of fertilizers and irrigation. Less than 30% of cultivated land is irrigated; thus agriculture is heavily influenced by monsoon patterns. Most Indian farmers harvest one crop per year, although double or even triple cropping is becoming more common in some areas.

India has very low levels of agricultural productivity, and average landholdings are small. Despite land reform limiting the size of large holdings, semifeudal land tenure persists in some parts: the most recent figures available indicate that 10% of households hold 53% of cultivated land, just 1% more than held the same acreage in 1954/55. Most small farmers have fragmented landholdings and poor access to credit and modern agricultural inputs.

Agricultural output is heavily influenced by the annual monsoon. The main foodgrain crops (the *kharif* crop—predominantly rice) and some cash crops (oilseeds, cotton, jute, and sugar) depend on the southwest monsoon, which brings 80% of India's rain, usually in a three-month period from June to mid-September. A second, northeast, monsoon brings lighter rains to the south from mid-October to December, when a winter (*rabi*) crop of wheat and coarse grains in grown in the north.

Parts of Indian agriculture have experienced the Green Revolution over the past three decades. High-yielding varieties have been introduced, combined with the timely use of inputs such as fertilizer and irrigation. But the increases in

---

This chapter is based on material prepared by Hiru Punwani and Hema Viswanathan except for "Overview of the Agricultural Sector", which is excerpted with permission from Economist Intelligence Unit, India, Nepal: Country Profile 1998–99 *(London: 1998), with minor stylistic alterations.*

output have not been evenly spread, and larger landholders are better able to afford essential inputs and additional labour. Small average landholdings, overintensive cropping, inappropriate use of fertilizers, and insufficient reach of irrigation facilities have also limited potential gains.

Agricultural policy has been dominated by a drive for self-sufficiency, especially in foodgrains. Since the mid-1970s India has maintained a comfortable buffer stock to protect against failed harvests and seasonal fluctuations.

But a policy of self-sufficiency has played havoc with cash crops. The production of and trade in sugar, cotton, jute, vegetable oil, tea, and coffee imports and exports have been managed to meet domestic consumption first. As a result, exports have been erratic and imports have often been misjudged, causing periods of shortages and surplus.

Target Crop and Audiences

Cotton accounts for about 60% of the plant protection products used in India because it is highly susceptible to pests, especially insects. Cotton is sprayed with several rounds of pesticides during the season, which in the case of high-yielding varieties lasts for about five months. Consequently, farmers growing cotton are much more exposed to pesticides than other farmers. Hence cotton was chosen as the target crop for the project.

The state of Tamil Nadu, in South India, has a large number of small-scale farmers who cultivate cotton but with inadequate knowledge of the safe and effective use of plant protection products and inappropriate practices. Moreover, the area also has many contract spraymen who are hired by farmers to spray their crops throughout the season. These individuals often spray pesticides throughout the day for five to six months at a stretch, thereby exposing themselves to a high degree of contamination. Our selected target audiences were therefore small farmers planting cotton (less than 2 hectares) and contract spraymen.

Target Area

The geographical area chosen for the project in India was the Coimbatore district. The district has a total of 480 villages. The intervention programme was tested in the Udumalpet block, and the control area was defined as the rest of the Coimbatore district. Udumalpet block has 97 villages, 65 of which traditionally cultivate cotton. From year to year, however, some farmers switch from cotton to non-cotton crops (such as coconut, bengal black gram, groundnut, and vegetables) and vice versa because of inadequate rainfall or poor or higher prices for cotton.

The cotton season commences normally in August or September with the onset of the southwest monsoon and continues until January or February in the following year. From October to December, the Coimbatore district receives rain mainly from the northeast monsoon.

The prevalent situation in the district was one of relative economic well-being in recent years. As a result, the second generation was being given the opportunity for education, which often resulted in their aspiring to a life away from agriculture. This led to an aging farmer population, many of whom saw merit in turning their land over to coconut plantations rather than agricultural crops. On the other hand, marginal farmers who had the opportunity sold their land to land development agents. Thus the farmer who will eventually stay with cash crops will be the professional farmer, a manager of agriculture. While the first steps in this direction may have been taken, it is a slow process before reaching the point where cash crops are no longer grown by small and marginal farmers.

Work on the farm is rarely a one-person job. More often than not, younger family members as well as wives play a helping role. This had an impact on the spraying operation when the mixing of the spray solution, the pouring of this solution into the spray-tank, and the spraying itself involved more than one person.

The economic, political, and climatic conditions during the seven years of the study were quite stable. The economic liberalization sweeping through India did not seem to have a direct impact on the lives of the Coimbatore cotton farmer, but there were years in which the cotton crop fetched good prices and the rains were good, which had an overall positive impact. The increasing reach of television, including the advent of cable television network in some villages, led to greater exposure to urban life-styles, which in turn fuelled ambitions for economic prosperity. The need for effectiveness from pesticides had been high in this region to begin with. This remained, in the final analysis, an important requirement.

Cultural Characteristics of Target and Control Populations

The farmers of Coimbatore District have long held strong beliefs on soil management, seeds and sowing, the use of manure, crop protection, and other aspects of farming. And many of them have proved to be rational ones, according to research done in the area.

Farmers believe, for example, a rainbow in the eastern sky brings heavy rain, and that when dragonflies fly low it may rain, whereas if they fly high, it is less likely to rain. They believe red soil is suitable for continuous cropping, and that deeply ploughed gardenlands conserve more moisture. Tilling the soil in April ensures a good crop, they maintain, through moisture conservation and control of pests.

Using organic manure gives you better crops, according to local farmers. Concerning pest control, they believe that continuous drizzling or a strong wind at the end of December increases the incidence of pests. And that it is important to fence a field before sowing in order to avoid damage by animals and birds.

These are all rational beliefs, according to scientists. In fact, out of a total of 62 traditional beliefs identified in one study, 83% were considered rational. And 87% were reported as strongly held.

Baseline KAP Study

In 1992, focus group discussions were conducted with farmers and with spraymen. The discussions were exploratory in nature and helped us design the project by providing an understanding of the range of knowledge, attitudes, and practices (KAP) that existed and that the project would need to address. (Focus group sessions were also conducted in 1994, after the first year of intervention, to see if adjustments of the media mix and contents were needed and, if so, which aspects had to be adjusted.)

The baseline study consisted of 6–10 group discussions with typically 8–10 farmers or spraymen in each, guided by a trained and experienced focus group moderator from Social and Rural Research Institute, which is a specialist unit of the Indian Market Research Bureau (IMRB). In order to assure valuable discussions and comfort levels within the group, separate discussions were held across different age bands and for farmers and spraymen. Structured questionnaires were administered in September through December 1992 by trained and experienced field investigators, under the supervision of IMRB field executives.

The interviews were conducted in the local language, Tamil. Although the questionnaire was printed in English, each interviewer carried the approved translation of each question. This ensured that no bias entered the study in the process of translation by an interviewer. It also maintained consistency across interviews and interviewers.

The focus group discussions established that cotton farmers in 1992 were alert to pests and considered them to be a major problem in farming. They believed that modern farmers had no choice but to use pesticides even though they blamed pesticides for the growing pest problem. Previous generations had been able to manage with biological pesticides such as neem leaves, farmers maintained, because there had been a lower infestation problem. Today, pest infestation had grown and, they conceded, indiscriminate use of pesticides had added to the problem. They believed that without pesticides, they could not expect high yields, which they wanted more than anything else.

Safety was not a major concern to farmers in 1992 because they did not see any significant examples of intoxication. Troubles such as headache, giddiness, and skin irritation were just accepted as occupational hazards. Farmers expected efficacy from pesticides and, when probed further, were concerned about the safety of the plants. Humans, they believed, had to fend for themselves and should look after their own safety by not spraying against the wind, by not allowing concentrate to spill on the skin, and by having a bath soon after completing the spraying job. It seemed that, at a subconscious level, farmers did not really believe in the health damage that could be caused by pesticides. Second, a poor yield could have immediate negative effects on farmers' economic situations and on the well-being of their families; poor health at a later date, on the other hand, could be dealt with when it occurred. There was no real recognition of long-term damage nor, it seemed, a desire to look deep into the problem because the alternative—limiting pesticide use—was unthinkable.

Intervention Programme

The intervention programme was launched in July 1994 in 58 villages of Udamalpet block, when nearly 500 hectares were planted with cotton. Project management decided to run the communications and training programme over three cotton seasons—1994/95, 1995/96, and 1996/97—because of the annual shift from cotton to other crops and vice versa by some farmers. This would ensure that all cotton growers were exposed to the prógramme. (See Table 4.1.)

Table 4.1. India: Area and Farmers Involved in Cotton Growing in Udamalpet

Cotton season	Villages planting cotton (number)	Farmers growing cotton (number)	Area covered with cotton (hectares)
1994/95	58	612	493
1995/96	62	1,201	814
1996/97	54	775	769

Basic Strategy

As noted earlier, the baseline KAP research had indicated that farmers were more concerned with maximizing their economic returns than with protecting their health and safety when using pesticides. The contract spraymen, however, were far more aware of health and safety issues and obviously did not worry about the economic aspects. The project management therefore decided to persuade farmers to adopt recommendations regarding safe use of pesticides by first demonstrating the economic benefits they would derive by following our recommendations regarding effectiveness. Hence the slogan used throughout the training programme was:

Protect your crop whilst you protect yourself.

Six key messages were emphasized in various media:

- Use the right pesticide at the right dosage at the right time.
- Wear gloves and use stick to mix pesticide.

- Wear protective clothing before starting to spray.
- After finishing spraying at the end of the day, take a bath with soap.
- After taking a bath, wash the protective clothing with a detergent.
- Clean the sparkplug and the gate valve of the mist blower regularly after finishing spraying at the end of the day.

Training Programmes

Project management hired four technical assistants for the project area, and a certain geographic territory was assigned to each person. These technical assistants were intensively trained by a full-time project leader who was in overall charge of the effective implementation of the programme. A local Steering Committee was set up for the project that included current and former Novartis personnel as well as two professors from Tamil Nadu Agriculture University.

The role of the local Steering Committee was to decide the broad parameters of the intervention programme after analysing the conclusions of the quantitative and qualitative KAP studies, to monitor the implementation of the intervention programme and decide on any corrective actions, and to make appropriate recommendations to the International Steering Committee.

Farmers' Meetings

The core of the intervention effort was a comprehensive programme of educating farmers and contract spraymen on the safe and effective use of plant protection products. (See Table 4.2.) This was accomplished through special training programmes in different parts of the project area, giving the invitees about one week's notice. In addition, impromptu, on-the-spot group meetings were convened by the technical assistants in the course of their daily extension activities. The education was imparted in an interactive manner, using aids such as flip charts, posters, handouts, and slide shows as well as a 20-minute film specially produced for the project. The film was both educational as well as entertaining, with a story woven around the main messages on safety and effectiveness of plant protection in cotton, giving it a local flavour and ambience. In the second and third year of the programme, the film became available to a wider audience through cable television.

A detailed and illustrated diary was given to each farmer. This elaborated the agricultural practices and plant protection measures for cotton cultivation with a provision for recording the pests identified, the pesticides actually sprayed in each round, the costs and expenses incurred, and the yield obtained.

With a view to initiating integrated pest management, farmers were educated to scout their crop and to assess the level of pest infestation, in a simplified manner. They were asked to commence spraying pesticides only after the incidence of infestation reached the economic threshold level (ETL). It was

Table 4.2. Summary of India Intervention Programme

Component	Number of times			Number of people attending/viewing		
	1994/95	1995/96	1996/97	1994/95	1995/96	1996/97
Farmer training programmes	63	62	61	4,798	3,435	3,660
Group meetings of farmers	80	72	120	880	350	1,080
Training contract spraymen	3	3	7	94	120	115
Campaigns for sprayer maintenance	8	5	7	150	63	110
Film on cable TV		78	314		2,340	10,048
School programmes	7	18	18	3,396	3,527	7,314
Contacts with doctors				44	45	45
Training extension personnel	2	2	1	126	105	64
Dealers' meetings	1	1		34	32	
Field days	1	2	2	414	510	612

observed that farmers who had been exposed to our educational programmes for two consecutive crop seasons began taking ETL-based decisions regarding spraying. They reported that they were thus able to reduce the number of spray rounds and the cost of plant protection measures significantly without adversely affecting yield. A film explaining the implementation of the ETL concept in a practical manner was screened during the training programmes for farmers.

During the training programmes, quiz sessions were held to check the degree of comprehension as well as retention of the six key messages of the programme. Useful prizes, such as spark plugs for sprayers, were distributed to those with the correct answer. This helped sustain the interest in the meetings and encouraged farmers to attend more than once. The key messages were also illustrated in coloured pictograms (for the benefit of the illiterate farmers and spraymen) with a short text in Tamil at the foot of each pictogram; these were printed on one sheet of paper that was distributed as a handbill during the 1996/97 crop season.

Protective Clothing

An important component of the educational effort involved recommendations on the use of protective gear by farmers and contract spraymen. The gear consisted of a cap, face shield, hand gloves, cotton shirt, cotton trousers, and a plastic apron, in accordance with the recommendation of the Indian Association of Basic Manufacturers of Pesticides. Gloves were strongly recommended for preparation of the spray solution, whereas the remaining items were to be put on during spraying. Not only was the use of protective clothing and the rationale for it emphasized at the farmer training sessions, but actual use in the field was monitored continuously. To encourage use, prizes or gifts were given to farmers and spraymen who were wearing protective clothing when applying pesticides during surprise visits or spot checks.

A set of protective gear (normally costing 275–300 rupees) was sold at the heavily subsidized price of 60 rupees. A nominal sum was charged because experience had shown that when such items are given away free of charge, they are rarely used. During the three seasons of intervention, the project sold 2,500 sets of protective clothing and equipment at this subsidized rate. Individual items of protective clothing were also made available at commercial prices through dealers in the project area for farmers and spraymen who wished to replace wornout items.

Contract spraymen frequently reported that although it was uncomfortable to wear protective clothing in the warm weather generally prevailing throughout the cotton season, they nevertheless felt much better at the end of the day when they did so: they did not experience any headache, giddiness, or irritation in the eyes, as had so often been the case in the past. As a result, spraymen had reduced their visits to the hospital or clinic. Since farmers who sprayed only their own fields were not exposed to pesticides continuously (unlike the contract spraymen), they did not report the beneficial effects of protective gear to the same extent.

Spraying Techniques and Maintenance of Sprayers

To ensure proper distribution of pesticides on plants as well as safety during application, appropriate spraying techniques are essential. This was an integral component of the education package. It was observed, however, that the target audience also needed information on the proper maintenance of spraying equipment. Sprayers were often leaking, which was a safety hazard. Several parts of the sprayers were in urgent need of repair and replacement. So sprayer maintenance campaigns were organized in collaboration with the leading manufacturer of sprayers (American Spring and Pressing Works). Farmers were invited to bring their spraying equipment to these camps, where it was checked by the engineers of American Spring and suitable advice was given free of charge.

To demonstrate the degree of contamination that can result from improper spraying of pesticides, a "water-sensitive" paper was used. Small pieces of this paper were pinned to different parts of the farmer's or sprayman's clothing, and droplets of the spray solution falling on the paper were easily visible because the paper turned blue wherever the droplets fell. These demonstrations were held on field day visits to demo plots, and farmers readily appreciated the need to adopt safe spraying practices to minimize, if not eliminate, contamination.

Other Components of Intervention

Demo Plots

In all three cotton crop seasons, eight demonstration plots of one tenth of a hectare each were laid out in farmers' fields by the four technical assistants under the supervision of the project leader. The agricultural practices and the recommendations regarding crop protection contained in the diary provided to all farmers were adopted in each plot, in order to show what can be achieved by following the recommended measures.

Before spraying, the crop was inspected, pests were identified, and the degree of infestation was assessed. This provided the basis for deciding on the most suitable pesticide (not necessarily one produced by Novartis) and dosage. These products were then sprayed, following the spraying techniques and practices recommended.

A demonstration board was displayed at each plot detailing the pesticides sprayed in each round and their respective dosages. During the growing season, field days were organized at each demo plot to give farmers from various villages an opportunity to see the results in field conditions as well as to compare the condition of the demo plot with neighbouring fields. At the end of each season, yield and cost data were collected for each plot and the cost/benefit ratio was calculated for comparison with similar data obtained from the fields of adjacent farmers. In most cases, the demo plot results were far more favourable than those of nearby fields, thereby convincing the farmers of the effectiveness of the recommendations.

Programmes for Schoolchildren

Since children can exert a substantial influence on their parents regarding safe practices in many situations, programmes at schools were held by our technical assistants to educate children on the "Do's and Don'ts" of pesticide application. The programme included the use of coloured pictograms in poster format, like the charts printed and recommended by the Association of Basic Manufacturers of Pesticides. In view of the encouraging and enthusiastic response from the schoolchildren as well as the teaching faculty, during the 1996/97 crop season the children were taken by bus to the demonstration plots to give them a first-hand experience of field conditions. The school authorities recently recommended to the Government of Tamil Nadu that crop protection be included in the school curriculum.

Education of Extension Personnel

To supplement and reinforce the work with farmers, extension personnel of the State Agriculture department were trained to spread the various messages on safe and effective use of pesticides to farmers.

Contacts with Doctors

There were about 45 medical practitioners in the project area; they often received patients suffering from pesticides contamination, some of which were due to suicide attempts. A lower proportion were related to accidental or occupational poisoning. The doctors were individually contacted, and the exposure routes of different categories of pesticides and the antidotes were explained. Charts detailing the antidotes, printed by the Association of Basic Manufacturers of Pesticides, were given to each doctor for display in the dispensary.

Meetings of Pesticides Dealers

Pesticides dealers in the project area also played an important role in stocking and recommending suitable products to farmers. Special meetings of dealers were convened at the commencement of the first two seasons to tell them our objectives and key messages, and to ask for their co-operation in promoting our efforts.

Results of Intervention

Unfortunately, interpretation of the results of the intervention programme became complicated by a simultaneous agricultural extension programme that was launched in the control area by Tamil Nadu University. With the control and test areas being contiguous, the spillover effect of our own intervention and the extension programme may have worked both ways. This confounding variable may lie behind some of the apparently anomalous findings.

Qualitative Evaluation

As a followup to the initial discussion held with farmers in 1992, focus group discussions took place in 1996. As with the quantitative surveys, these were held in both the intervention and control areas in order to be able to compare the developments with the programme to those without it. (As noted, it turned out that this analysis became difficult because of the simultaneous agricultural extension programme in the control area.) The 1996 sessions were diagnostic in nature and served to explain why some recommended practices had been adopted while others had not. (Unlike the quantitative surveys, the qualitative ones were not repeated one year after conclusion of the intervention.)

Differences Between the Intervention and Control Areas in 1996

There was clearly a higher level of knowledge among test area farmers about the need for safety in the way pesticides were handled. Those who had been exposed to several different education media used by the project (the television film, the calendar, the demo plots, and the farmers' meetings) showed high awareness of safety measures. Such measures often fell into disuse after cessation of the intervention, however, because of their inconvenience. In contrast, there was a lower level of awareness and concern in the control area.

The control area farmer had also learned about some protective measures, but was not able to trace this directly to a learning occasion. Farmers spoke of sessions that had been conducted by companies that produced competing brands of pesticides and by agriculture extension workers. But they also said that they had not attended all such sessions. Some also spoke of advice being given by dealers.

Project managers believe that there was some spillover of learning from the intervention region to the control area. Some of the information could have been picked up and adapted by the agricultural extension workers who moved freely between the two areas. Some of it could have been provided by other pesticide companies in an effort to keep pace with the Novartis work. Finally, there was considerable social and business interaction between farmers of the two regions and their families, and some learning could have been exchanged in this process.

Differences Between 1992 and 1996 in the Intervention Area

Farmers in the project region seemed to have understood and internalized the need for the use of protective gear. They accepted that there was need for protection, and many had adopted a full-sleeved shirt and full trousers as an appropriate spraying outfit. This offered substantially more protection than the "half-lungi" that was worn in 1992.

Other protective measures, such as the use of gloves, did not get adopted as well, however. This was partly explained by the inconvenience of gloves, but it was also explained by the fact that the entire process of measuring, mixing, and pouring the diluted liquid in the spray-tank involved different people, often over different repeated mixing and pouring sessions. The logistical difficulties of the exercise as well as the inconvenience made the adoption of gloves less likely.

Similarly, there was improvement in knowledge in such areas as the indication of toxicity on the pack or the existence of beneficial insects. This did not seem to translate into action, however, in that there appeared to be little understanding of the use that could be made of the information.

Finally, there were clear changes in attitude that reflected a recognition of the fact that unsafe practices could damage health and that a farmer's health was not unimportant. This was an improvement over 1992, when farmers believed that their health was not as important as a good harvest and economic gain. Now farmers recognized that their health could be affected and was not to be neglected. The attitude of self-neglect had been replaced to some extent by a recognition of self-worth.

Farmers' Perception of the Changes and the Project

There was widespread recognition among farmers that Novartis had taken a lot of care to improve farmers' health by teaching them safety methods and to improve their yields by teaching about the effective use of pesticides. There was appreciation of the fact that no attempt had been made to sell the company's products through this project. This added to its credibility and appeared to have created good will towards Novartis personnel working in the region.

The protective gear sold in the project region at a subsidized rate created a lot of interest. However, it appeared that with the cessation of the project and the resulting nonavailability of the gear at a subsidy, interest waned. This suggests that the experience of the benefits of protective clothing were not carried out long enough to have become internalized. A combination of poverty and an in-built sense of thrift made farmers look for low-cost alternatives, which they found in wearing full shirts and trousers.

The film was well recalled and appreciated. Though not directly mentioned, informal feedback suggested that the training programme for children at school played an important role in bringing the messages home. The farmers' meetings and demonstration plots gave a serious and technical angle to the entire project, which appealed to the farmers and made them take the project seriously. Finally, the communications of the Novartis Project team made a positive and significant impact, adding to the sense of importance and gravity of the project.

Quantitative Evaluation

Methodology

In keeping with the theory of measuring change under controlled conditions, this study was designed as a longitudinal, cross-sectional comparative study. Thus there was the project region in Udamalpet and a control region in the rest of Coimbatore district. A cross-sectional study among a randomly selected sample of farmers and spraymen in both regions was carried out over three different time periods—at the baseline in 1992, at the endline study in 1996, and in a follow-up study in 1997.

As described in Chapter 3, the programme effect should be seen in the difference between the changes in the project and the control area, or "net change". There are two levels of net change. The first is the change between the 1992 baseline and the 1996 endline (Net); the second is the change between the 1996 endline study and the 1997 follow-up study (Net+). Net quantifies the proportion of test area farmers who have adopted some knowledge, attitude, or practice as a result of the project, while Net+ quantifies the persistence of adoption. A positive Net+ indicates that adoption in the test area increased after the end of the project (in comparison with the control area), possibly because communication between farmers continues to spread the particular knowledge, attitudes, or practices. A Net+ value near 0 indicates that adoption stayed approximately at the level achieved when the intervention stopped, and a negative Net+ value indicates that adoption decreased after the end of the project (always in comparison with the respective control area development).

One disadvantage of this approach is that it assumes that the outside agents that affected the control area were as active in the project area. However, this assumption may not always be valid. Certain changes may have taken place exclusively in the control region, as, for example, the additional inputs given by the agricultural extension workers, based on the teachings that they saw in the project region. Such inputs can skew the results of the "net change" model. Therefore the absolute changes in the test area, without consideration of the control area, should be and are also analysed.

Media Reach

In 1996, the 511 farmers who were interviewed in the project area and the 508 farmers interviewed in the control area were asked if they had received any information or advice on pesticide usage. In the project area, 73% said they had received information; in the control region, 25% reported this. The posters and talks at farmers' meetings were the main sources mentioned. (See Table 4.3.)

Table 4.3. India: Sources of Information on Pesticide Usage

Source	Project area	Control area
Film/video show	58	13
Poster	77	7
Talk and slide/chart show	78	15
Diary/book	44	7
Calendar	48	7
Demo plots	50	1

Of the farmers who were able to recall a communication, 39% identified Novartis as the source.

Adoption of New Attitudes and Behaviour

Tables 4.4 to 4.8 provide overviews of the values at different times of a number of variables assessed with the KAP and observation surveys. All variables are defined such that the desired changes are positive. For example, instead of talking about the proportion of farmers who had minors mix pesticide solutions, the project looked at the proportion who did not have such mixing done by minors. It is important to remember that Net+ indicates the change between the endline and follow-up surveys, not between the original baseline survey and the follow-up.

Summary of Major Findings

A comparison of the impact in the project region versus the situation in the control region is necessary, together with an analysis of the net impact in the region. This section looks at the changes that took place over the years in the project region and examines these in the context of changes that might have taken place over the same period in the control region. All the changes reported are statistically significant.

Continued Changes Due to the Project

In a number of cases, the change in knowledge, attitudes, or practices was in the desired direction and continued even after the communication and training had been stopped for a year.

- Condition of Spraying Equipment

In 1992, 84% of the farmers in the project region reported that their spraying equipment was in perfect working condition. This went up to 93% in 1996 and then to 97% in 1997. In the control region, 85% and 88% reported equipment in good working condition in 1992 and 1996 respectively. This rose to 96% in 1997. The significant improvement in both regions in 1997 over 1996 points to some intervention outside the project in that one year. The trend was similar for spraymen.

- Determining Volume of Pesticide to Buy

Farmers in the project region who used the size of the plant or the extent of the problem or both to determine how much pesticide to buy rose from 10% in 1992 to 61% in 1996, and then to 70% in 1997. This learning had clearly been sustained. However, this was also true in the control region, where the relevant response rose from 6% in 1992 to 85% in 1996 and 82% in 1997. Thus although clearly there was some learning involved, it cannot only be attributed to the project.

Table 4.4. India: KAP Survey Results on Farmers' Knowledge (percentage)

Knowledge	Test area			Control area			Test area changes		Net changes	
	B	E	F	B	E	F	E-B	F-E	Net	Net+
Face mask[a]	73	82	62	66	76	53	9	-20	-1	3
Overalls[a]	20	26	21	17	3	8	6	-5	20	-10
Gloves at mixing[a]	0	5	7	0	19	3	5	2	-14	18
Full trousers[a]	21	73	50	11	28	19	52	-23	35	-14
Full shirt[a]	21	81	59	11	44	25	60	-22	27	-3
Beneficials[b]	21	64	44	21	47	42	43	-20	17	-15

B = baseline E = endline F = follow-up

[a] Farmer mentions specific protection when asked about ideal protective measures to take when using or mixing the most toxic pesticide he has ever used.

[b] Farmer mentions that some insects are beneficial rather than being pests, on a direct question.

Table 4.5. India: KAP Survey Results on Farmers' Attitudes (percentage)

Attitude	Test area			Control area			Test area changes		Net changes	
	B	E	F	B	E	F	E-B	F-E	Net	Net+
Health main problem[a]	18	25	30	22	40	27	7	5	-11	18
Health risk exists[b]	77	76	61	66	84	68	-1	-15	-19	1
No risk, precautions taken[b]	25	2	22	32	14	6	-23	20	-5	28
Not fatalistic[d]	18	23	16	14	26	21	5	-7	-7	-2

B = baseline E = endline F = follow-up

[a] Farmer mentions health problems from spraying when asked which are the main problems associated with pesticide use.

[b] Farmer answers yes on the direct question of whether a health risk is associated with the pesticide spraying job.

[c] Farmer reports following all necessary safety precautions when asked why he believes there is no health risk associated with the spraying job, expressed as proportion of those indicating that there is no such risk.

[d] Respondent disagrees with statement "Man has to take risks. Spraying is my risk, I do not mind."

Table 4.6. India: KAP Survey Results on Farmers' Reported Practices (percentage)

Practice	Test area			Control area			Test area changes		Net changes	
	B	E	F	B	E	F	E-B	F-E	Net	Net+
Checked pest[a]	88	99	79	84	98	93	11	-20	-3	-15
Scouting scheme[b]	48	67	63	46	36	63	19	-4	29	-31
Wash before eating[c]	74	87	68	56	58	83	13	-19	11	-44
Face mask[d]	29	62	27	28	56	17	33	-35	5	4
Overalls[e]	1	32	11	0	1	1	31	-21	30	-21
Full trousers[e]	14	47	20	6	4	2	33	-27	35	-25
Full shirt[e]	14	67	38	6	25	12	53	-29	34	-16
Cleans sprayer with water[f]	50	50	22	47	51	38	0	-28	-4	-15

B = baseline E = endline F = follow-up

[a] When asked how he decided to use a pesticide, farmer answered "by checking pests".

[b] Farmer reports using a fixed scouting scheme, as opposed to "eyeball survey" or not inspecting the crop.

[c] Farmer reports taking precautions before eating, drinking, or smoking after spray work. (Further analysis shows that these precautions are washing the hands and, in few cases, the body.)

[d] Farmer reports using a face mask or cover when asked what precautions taken while spraying (last time).

[e] Farmer reports using specified protection when asked what precautions taken while spraying.

[f] Farmer reports cleaning spraying equipment by pouring water in the tank and spraying it out.

Table 4.7. India: KAP Survey Results on Farmers' Health (percentage)

Health indicator	Test area			Control area			Test area changes		Net changes	
	B	E	F	B	E	F	E-B	F-E	Net	Net+
Never health problem[a]	76	54	61	77	63	58	-22	7	-8	12
Skin problem, ever[b]	4	11	2	4	14	10	7	-9	-3	-5
Headache, ever[b]	1	5	9	4	16	16	4	4	-8	4
Giddiness, ever[b]	4	3	3	8	6	3	-1	0	1	3
Nausea, ever[b]	1	0	1	0	1	1	-1	1	-2	1
Vomiting, ever[b]	1	2	1	1	1	1	1	-1	1	-1
No health problem last time[c]	69	57	46	57	31	46	-12	-11	14	-26
Did nothing about health problem[d]	18	31	40	7	25	29	13	9	-5	5

B = baseline E = endline F = follow-up

[a] Farmer answers no when asked whether he has ever suffered a health problem due to pesticides.

[b] Particular problem mentioned by farmer when asked about main health problems due to pesticides, expressed as proportion of those who reported such problems.

[c] Farmer's answer does not specify any symptom when asked whether he has suffered a health problem due to pesticide on the latest spraying occasion.

[d] Farmer did nothing about a health problem.

Table 4.8. India: Observation Survey Results on Farmers' Practices (percentage)

Practice	Test area			Control area			Test area changes		Net changes	
	B	E	F	B	E	F	E-B	F-E	Net	Net+
Face mask[a]	n.a.	15	3	n.a.	4	2	n.a.	-12	n.a.	-10
Towel[a]	n.a.	32	37	n.a.	57	63	n.a.	5	n.a.	-1
Full trousers[a]	n.a.	21	4	n.a.	0	1	n.a.	-17	n.a.	-18
Full shirt[a]	n.a.	41	14	n.a.	14	11	n.a.	-27	n.a.	-24
No sprayer leaks noted	62	96	95	80	98	92	34	-1	16	5

B = baseline E = endline F = follow-up

n.a. = not available

[a] Observed attire while spraying.

- Colours as Indication of High Toxicity

Of all farmers interviewed in the project area, the proportion who could correctly identify red as the colour for the most toxic pesticide changed from 7% in 1992 to 11% in 1997. In the control region, 11% in 1992 could correctly identify red as the colour indicating the most toxic pesticide. This dropped to 7% by 1997.

- Recommended Dosage as a Consideration for Concentrate:Water Ratio

Only 5% of the farmers had mentioned this in 1992. This rose to 36% in 1996 and stayed persistent at 30%. In the control region, 10% of farmers mentioned this consideration in 1992, 62% in 1996, and 29% in 1997.

- Use of a Measuring Glass by Farmers

This practice was at a high level in 1992, at 91%. The proportion fell in 1996 to 83% but rose again upon cessation of communication to 92%. On the whole, therefore, there was no change on this issue. However, the practice declined in the control region quite dramatically—from 83% in 1992 to 50% in 1996 and 29% in 1997. As a result, there is a clear and positive net change, which suggests the intervention programme succeeded in ensuring that a good practice was continued when the surrounding ambience was shifting away from this practice. In a changing situation, in other words, a positive practice was sustained.

This was further borne out by observation, during which a steady increase was found in the use of a measuring glass for measuring pesticides. From 64% who followed this practice in 1992, the proportion rose to 72% in 1996 and further to 86% in 1997.

Among spraymen, it was found that the practice of using a measuring glass fell in the project region (from 95% in 1992 to 78% in 1996) but rose again towards 1997 (to 89%). But there was a continued fall in the control region (from 84% in 1992 to 84% in 1996 to 17% in 1997). Thus there was a positive net change on this issue among spraymen too.

- Use of Bare Hands for Mixing Pesticides

This practice was high among farmers in 1992, at 85%. It fell to 63% in 1996 and fell further in 1997 to 29%. Thus, there was a steady decline in this unfavourable practice. In parallel, in the control region, the use of bare hands dropped from 74% in 1992 to 61% in 1996, but it rose again in 1997 to 69%. Thus there was apparently some learning in the control region, but it was not sustained. This change was also found among spraymen in the project region, where use of bare hands fell from 86% in 1992 to 41% in 1996 to 31% in 1997.

- Knowledge on Use of Gloves While Mixing Highly Toxic Pesticides

Knowledge regarding the need to use gloves while mixing highly toxic pesticides went from nil to 5% among project region farmers and further increased to 7% by 1997. In contrast, in the control region, knowledge regarding the need to use gloves while mixing toxic pesticides went up sharply—from nil to 19% in 1996, but this was not sustained. In 1997, only 3% of the farmers in the control region mentioned the need.

Among spraymen in the project region, awareness rose from 2% in 1992 to 8% after the communication programme, but fell to 4% on cessation of communication.

- Farmers' Practice of Not Taking Breaks for Food, Drink, or a Cigarette

This practice was relatively high in 1992, with 68% of all farmers not taking breaks. This rose to 84% in 1996 and was sustained at 85% in 1997. In the control region, the practice remained fairly steady at 74% and 77% between 1992 and 1996, but rose sharply in 1997, with 87% of the farmers saying that they had not taken any breaks this time while spraying, suggesting that there had been some outside teaching in 1996–97. A visit to the control area uncovered several inputs during the year from the agricultural extension workers as well as from the agricultural universities and other crop protection companies; farmers had been told that they should not spray for more than four hours a day. This explains the drop in the felt need for a break.

In the case of spraymen, the proportion who did not take breaks rose in 1996 over 1992. But there was a sharp drop in this in 1997, probably because it was impractical for men who sprayed for several hours on end in the season to do so without a break.

- Taking Precautions While Taking Breaks

Spraymen and farmers who did take breaks took care to take precautions before eating, drinking, or smoking. These precautions usually took the form of washing hands, gargling, and washing the face. The proportion of farmers and spraymen who took food or drink or smoked during breaks without taking any precautions was very small, hovering between 0% and 5% of all respondents in all years in both project and control regions.

- Use of Full Trousers and Shirt While Spraying

In the project region, use of a full-sleeved shirt and/or full trousers was at the low level of 14% in 1992. This rose significantly after the project intervention. The use of full-sleeved shirts went up to 67%, while the use of full trousers went up to 47%. After cessation of communication activities, this proportion fell to 38% and 20% respectively. Thus, while there was a drop after the campaign, the overall change was still positive. This was a positive net change, particularly in the case of full trousers.

By contrast, in the control region, use of full shirts went up from 6% in 1992 to 25% in 1996 but fell back to 12% in 1997. Thus the learning was not sustained. The use of full trousers, on the other hand, did not register in the control region at all. From a low base of 6%, their use fell to 4% in the control region in 1996 and then to 2% in 1997.

In the case of spraymen, the change was similar. The use of a full-sleeved shirt and/or full trousers was at a low level of 7% in 1992. This rose significantly after the project intervention. The use of full-sleeved shirts went up to 64%, while the use of full trousers went up to 38%. After communication activities stopped, these proportions fell to 46% and 18% respectively. Thus although there was a drop after the campaign, the overall change was still positive. In the control region, the use of full shirts continued at levels that were better than in 1992, but that showed a sharp drop over 1996. The use of full trousers showed slow growth, but was still practiced by a very small proportion of spraymen.

- Farmers' Practice of Having a Full Body Wash After Spraying

In the project region, the practice of having a full body wash after spraying stood at 68% among farmers in 1992. This rose after the project intervention to 88%. And after communication activities ended, it reached 90%. Thus, there was a positive change in the practice and this persisted such that the overall change was positive. At the same time, the proportions in the control region who followed this practice were 67%, 95%, and 95% respectively. This takes the credit away from the project as the practice could have improved due to other factors; alternatively, there could have been a spillover of the learning from the project region.

- Washing Clothes Worn While Spraying

Though no 1992 data are available for this, the study found that 87% of all farmers in 1996 followed the practice of washing work clothes in soap or detergent. In 1997, this practice persisted at 81%. In contrast, following this practice in the control region fell from 84% in 1996 to 55% in 1997. This suggests that the teaching in the test region played a role in keeping the practice at a high level.

- Farmers' Experience of Health Problems from Spraying

The proportion of farmers who had never experienced any health problems from spraying was 76% in 1992. This fell to 54% in 1996, but rose again to 61% in 1997. During the same time, the proportions in the control region who did not experience any health problems were 77%, 63%, and 57% respectively, showing improving health and less trouble than was found in the project region.

- Knowledge of Beneficial Insects

The proportion of farmers who had been aware that some pests are beneficial stood at 21% in 1992. This rose to 64% in 1996, after the project intervention. A year later, however, the knowledge level fell back to 44%. In the control region, over the same time period, the awareness shifted from 21% in 1992 to 47% in 1996 and 42% in 1997.

The same was true of spraymen, where the proportion who were aware of beneficial insects rose from 16% in 1992 to 48% in 1996, after the project intervention. A year later, this awareness had fallen to 39%. Thus the change had been sustained, albeit at lower levels than after the project intervention had taken place. Among control region spraymen, the comparable proportions were 23%, 54%, and 37%, showing a similar rise and fall in awareness.

- Attitudinal Change

Attitudes had been ascertained using a set of statements. Responses to these show that in the case of most attitudes, there has been a sustained positive change over time. (See Table 4.9.)

Table 4.9. India: KAP Survey Results on Attitudes and Perceptions (percentage)

Statement disagreed with	Project region 1992	Project region 1996	Project region 1997	Significant change over 1992 at 99% confidence level	Control region 1992	Control region 1996	Control region 1997
"Pesticides nowadays are too weak to be poisonous for humans."							
Farmers	25	52	38	Yes	19	22	44
Spraymen	33	54	54	Yes			
"When I spray, I feel like a general who fights enemies."							
Farmers	3	20	16	Yes	5	28	22
Spraymen	5	12	11	Yes, at 95% level			
"These masks and plastic covers while spraying are unmanly."							
Farmers	65	86	80	Yes	58	69	85
Spraymen	70	92	83	Yes, at 95% level			
"Man has to take risks, spraying is my risk, I do not mind."							
Farmers	18	23	16	No	14	26	21
Spraymen	17	32	13	No			
"A person who sprays, definitely harms his health in the long run, no matter how careful he is."							
Farmers	10	15	17	Yes	11	14	13
Spraymen	5	5	11	Yes, at 95% level			
"Spray persons have become immune to the effect of pesticides."							
Farmers	19	57	58	Yes	12	32	44
Spraymen	18	48	67	Yes			
"I am convinced that today we treat plants too violently; if we treated them more tenderly they would grow better." [N.B. proportion who agreed with statement]							
Farmers	71	90	81	Yes	77	85	25
Spraymen	75	100	66	No			

[a] Survey base in project region as follows: in 1992, 605 farmers and 240 spraymen; in 1996, 511 farmers and 50 spraymen; and in 1997, 498 farmers and 100 spraymen.
[b] Survey base in control region as follows: in 1992, 466 farmers; in 1996, 508 farmers; and in 1997, 514 farmers.

Desired Changes That Did Not Continue

In some cases, the change in knowledge, attitudes, or practices was in the desired direction but it did not continue after the communication and training programme stopped.

- Inspection of Plants Before Deciding to Spray

The proportion of farmers who inspected the plants changed from 88% in the project region in 1992 to 99% in 1996, but then dropped to 79% by 1997. A similar trend was seen in the control region, although the change there was more sustained in the positive direction than in the project region (84%, 98%, and 93% respectively).

- Deciding Which Pesticide to Use

In 1992, 79% of all farmers had taken a decision regarding the correct pesticide for their pest problem after checking for pests, and thereafter asking for advice or directly taking a decision. This correct practice rose to 82% in 1996, but fell back to 79% in 1997. Thus sustained learning was not recorded on this issue in the project region. In the control region, the correct practice was found among 84% of farmers in 1992. This remained at 84% in 1996, and then rose to 86% in 1997. The changes in the project region appeared, therefore, to be less satisfactory.

- Perception of the Importance of Safety

The proportion of farmers who ranked safety as the most important attribute needed from a pesticide went up from 8% in 1992 to 43% in 1996, but this was not sustained, as the importance of safety dropped to 19% in the 1997 study. In the control region, the proportion went up from 8% at the baseline (1992) to 21% in 1996 and then to 40% in 1997. Once again, this reflects the activities of other agencies in the control region in 1996–97.

The same was found among spraymen, where the proportion who ranked safety as the most important attribute in a pesticide went from 10% in 1992 to 60% in 1996, but came down to 37% in 1997. In the control region, the comparable proportions for spraymen over the years were 9%, 54%, and 54%.

- Method of Storing Unopened Bottles

The method of storing unopened bottles was already good in 1992 , when 97% had either stored the bottles in an acceptable manner or purchased the pesticide on the day it was used, thus avoiding storage altogether. This stayed at the same level in 1996, but not after the intervention programme ended. By 1997, the proportion had fallen to 94%. In the control region, the comparable proportions were 96%, 98%, and 94%. On the whole, however, the proportions who stored pesticides in an acceptable manner was high in both regions at all stages.

- Farmers' Measurement of Pesticides

The proportion who measured pesticides was already high in 1992, at 96%. This went up to 99% in 1996. After the communication efforts ceased, however, the proportion fell back to the 1992 level, at 95%. In the control region, the proportions were 81%, 87%, and 97% respectively, showing a steady increase in this practice. This makes the fall in the project region even more unsatisfactory.

- Use of Gloved Hands to Mix Pesticides

The use of gloves was nonexistent among farmers in 1992. With the communication programme, the practice rose to 17% in 1996. A year later, however, the practice fell back to 4%. The sharp nature of the fall suggests that the practice is not likely to be sustained. In the control region, the practice did not really get absorbed at all, moving from 0% in 1992 to 2% in 1996 and 3% in 1997.

Among spraymen, the use of gloves was reported by only 2–4% of all respondents between 1992, 1996, and 1997 in both control and project regions.

- Changing into Different Clothes for Spraying

This was found among 57% of the farmers in 1992, and rose to 69% after the project campaign. After the programme ended, however, this fell again to 42%. Nevertheless, the situation was still better than in the control region, where the practice initially rose from a good level of 45% in 1992 to 52% in 1996, but then dropped sharply in 1997 to 14%. Thus there was a negative picture in 1997 in both regions, but more so in the control area.

The same was true for spraymen, where the proportion in the project region who changed into different clothes for spraying rose from 65% in 1992 to 84% in 1996 and then fell back to 64% in 1997. In the control region, the proportion fell steadily from 54% in 1992 to 49% in 1996 and to 40% in 1997.

- Use of a Face Mask

The use of a face mask showed a sharp increase in the project region in 1996 (62%) over 1992 (29%); by 1997, however, the proportion had fallen back to baseline levels, with only 27% reporting the use of face masks. Thus this was rapidly learned but the change in behaviour did not continue. In the control region, the trend was very similar, with a 28% use reported at the baseline, which rose to 56% in 1996 but fell sharply to 17% in 1997.

Among spraymen, too, the pattern was similar—33% reported use of face masks in 1992, which rose to 50% in 1996 but fell sharply to 19% in 1997. In the control region, the proportions for spraymen were 32%, 35%, and 26%. Thus the learning in 1996 may be attributed to the project.

- Starting Spraying Operations Before 9:00 a.m.

Before the project intervention, 37% of the farmers had started spraying before 9.00 a.m. (which is preferable as it is cooler and therefore there is less evaporation, making pesticide use more effective). By 1996, this rose to 59%, but by 1997, the proportion had fallen back to 36%, resulting in no sustained change. In the control region, the proportion who started before 9:00 a.m. was 48%, 76%, and 57% respectively. Compared with the control region, therefore, the trend in the project region was not positive.

An identical pattern was found among spraymen, where the proportion who started spraying before 9:00 a.m. stood at 45% in 1992. By 1996, this rose to 72%, but by 1997, it had fallen back to 47%, resulting in no sustained change. In the control region, the proportions for spraymen were 51%, 75%, and 70% respectively over the three years.

- Cleaning of Spray-tank with Water from Inside

In 1992, 55% of the farmers had followed this practice during the most recent spraying occasion (with or without the use of soap). This rose slightly, to 57%, in 1996. A year later, however, the proportion who did this fell sharply to 27%. Thus this was a positive practice that fell back to below the original level. The picture was comparable in the control region, where reported cleaning of the inside of the spray-tank had been found at 54% in 1992. This rose to 56% in 1996, but dropped again to 44% in 1997. Observation showed that the actual levels of cleaning were in fact even lower—at 31%

in 1996 and 14% in 1997 in the project region. Comparable observation data for the control region showed 11% following this practice.

Of the spraymen, 64% had followed this practice in 1992 (with or without the use of soap). This remained steady in 1996, even as the practice showed a drop in the control region from 56% in 1992 to 35% in 1996. This indicates that the project played a role in holding the practice steady at 64% in the intervention area. Upon cessation of project activities, however, the proportion who followed this practice fell sharply to 32%, while it rose in the control region to 41%. Thus this was a positive practice that fell back below the 1992 levels in the project region.

- Frequency of Farmers Cleaning the Spray-tank

The spray-tank was cleaned every time it was used by 58% of the farmers in 1992. Though this proportion rose to 70% after the project intervention, it fell again to 58% a year after communication activities ended. Thus the change was not sustained. The control region also showed that 59% and 57% followed this practice in 1992 and 1997 respectively. There was a constant pattern regarding this practice that the project had not managed to affect.

No Change

In the following cases, there was no significant change in the attitudes or behaviour of farmers or spraymen as a result of the project:

- safe disposal of concentrate containers by spraymen,
- spraymen's practice of having a full body wash after spraying,
- considering the toxicity label of a pesticide package,
- recognition of health risk from spraying,
- recognition of who is at risk, and
- frequency of spraymen cleaning the spray-tank.

Changes in Undesirable Direction

In a half-dozen cases, the study registered changes in behaviour or attitudes that were not in the desired direction.

- Awareness that Expiry Date is Mentioned on Pack

Among farmers, this awareness fell from 71% in 1992 to 47% in 1997. In the control region, the awareness shifted from 71% in 1992 to 41% in 1997. (No data are available on this topic for 1996.)

- Awareness that Toxicity Level is Indicated on Pack

Among farmers, this awareness fell from 69% in 1992 to 32% in 1997. In the control region, it fell from 65% in 1992 to 33% in 1997. Among spraymen, too, a similar trend was found, with awareness falling from 55% in 1992 to 21% in 1997. (Again, no data are available for 1996.)

- "Extent of Infestation" as a Consideration for Concentrate:Water Ratio

In 1992 , 63% had taken this factor into consideration. This fell to 38% in 1996 and further to 20% in 1997. Both changes were significant at a 99% level of confidence. In the control region, the proportion who took this into consideration was 53% in 1992, 55% in 1996, and 5% in 1997.

- Farmers' Safe Disposal of Concentrate Containers

The practice of burying, breaking, or otherwise destroying the concentrate container was at a low level of 9% in 1992 among farmers. It fell to 8% in 1996 and then to 4% in 1997. A similar trend was found among control region farmers, where acceptable disposal methods fell from 9% in 1992 to 2% in 1996 and 1% in 1997.

- Recognition of Health Risk Each Time Proper Safety Precautions Not Taken

This recognition had been found among 44% of the farmers who believed there was a health risk at all in 1992. This fell to 13% after the project intervention. One year after cessation of activities, it fell further, to 7%. Thus an aspect of spraying on which there should have been increasing awareness actually fell over time. An even sharper drop was found among control region farmers who accepted that there was a health risk—from 38% to nil. The project managed to hold some who believed in this statement, but not a substantial number.

The same was true of spraymen: the proportion who gave this response fell from 38% in 1992 to 6% in 1996 and 9% in 1997. Among control region spraymen, the proportion fell from 34% to 1% between 1992 and 1997. (Again, no data are available for 1996.)

- Experience of Health Problems During Most Recent Spraying

The proportion who reported that they had experienced health problems during the most recent spraying episode rose steadily from 1992 to 1997. This is not the desired direction of the change. However, it could also be explained by a heightened awareness leading to higher reporting of problems.

Among farmers, 7% had reported one or more health problems during the recent spraying episode in 1992. This went up to 20% in 1996, and was sustained at 21% in 1997. In the control region, there were similar trends, with the proportion who had experienced health problems during the recent spraying occasion rising from 10% in 1992 to 26% in 1996 and further to 31% in 1997.

Among spraymen, 24% had reported one or more health problems during the recent spraying episode in 1992. This stayed at 24% in 1996, but rose to 43% in 1997. In the control region, the experience of health problems during the recent spraying episode was reported by 26%, 43%, and 46% of spraymen in 1992, 1996, and 1997.

Relationship Between Knowledge, Attitudes, and Practices

In terms of the knowledge base of those surveyed in the test area (see Table 4.10), farmers were already fairly knowledgeable at the start of the project about safe storage methods of concentrate containers and the need to use a face mask while spraying. Significant improvement was achieved on knowledge regarding appropriate doses, the need to use a full shirt and full trousers while spraying, the need to use gloves to mix pesticides, and the fact that some insects could be beneficial. But no real change was achieved in awareness regarding the need to use aprons.

Table 4.10. India: Overview of Knowledge on Pesticide Use in Project Region

Issue	1992	1996	1997
Deciding on the quantity to be bought on the basis of the plant size and/or the extent of the problem	10	61	70
Safe method of storing unopened bottles	90	94	90
Use of gloves to mix pesticides	0	5	7
Use of face mask (ideally)	73	82	62
Use of apron (ideally)	20	26	21
Full shirt	21	81	59
Full trousers	21	73	50
One or more of above four	79	91	78
Aware of beneficial insects	21	64	44

Significant improvement was also achieved in changing attitudes about the importance of safety as a pesticide attribute (although this did not continue), about the notion that pesticides were now too weak to be harmful, and about farmers' possible immunity to the harmful effects of pesticides. (See Table 4.11.) The campaign was also successful in removing inhibitions to the use of face masks and other protective gear on the grounds that these are unmanly. And there was a negative trend in the importance accorded to effectiveness; since the baseline study indicated a tendency to seek effectiveness at any cost, this negative trend is considered a desirable outcome. At the start of the project, farmers already recognized fairly well the risk involved in the spraying process, so there was not much change possible there.

No real change was achieved, however, in the recognition that other people present in the field during spraying could be at risk or in the attitude of acceptance with regard to risks. And there was a negative trend with regard to the belief about risk if a person did not take safety precautions. The smaller proportion who expressed this belief at the end of the project suggests that there was some confusion in understanding the role of protective measures in reducing risk.

In terms of stated practices among farmers in the project region, on three issues the occurrence was already at a high level, so there was not much change possible there. (See Table 4.12.) This was essentially with regard to practice of measuring pesticides before use and taking proper precautions during breaks.

Table 4.11. India: Overview of Attitudes on Pesticide Use in Project Region

Issue	1992	1996	1997
Importance of safety in pesticides	8	43	19
Importance of effectiveness in pesticides	88	52	77
Recognition of risk in the spraying operation	81	81	73
No risk if safety precautions followed	6	9	7
Risk when a person does not take precautions	44	13	7
Risk for others in the field too	45	50	46
Pesticides too weak to be harmful	25	50	38
Masks are unmanly	65	86	80
"This is my risk, I do not mind"	18	23	16
"We have become immune to the harmful effects"	19	57	58

Table 4.12. India: Overview of Stated Practices of Pesticide Use in Project Region

Issue	1992	1996	1997
Decision taken correctly regarding the need to use pesticides	88	99	79
Decision taken correctly regarding which pesticide to use	79	82	79
Used after measuring	96	99	95
Mixed using measuring glass	91	83	92
Mixed using gloves	0	17	4
Changed into different clothes for spraying	57	69	42
Took no breaks or took proper precautions	96	98	96
Reports of getting wet from spray, leaks, spills	-	59	63
Cleaned equipment each time	58	70	58
Disposed of empty containers correctly	9	8	4
Less than 30 minutes gap between spraying and washing	-	74	65
Washed spray clothes each time	-	93	86
Used face mask last time	29	62	27
Used apron last time	1	32	11
Used full shirt last time	14	67	38
Used full trousers last time	14	47	20

Significant improvement was achieved in practices pertaining to the correct methods of determining the need to use pesticides, in the use of gloves while mixing pesticides, and in the use of some protective gear (face masks, apron, full shirt, and full trousers) while spraying. No real change was achieved, however, in the practice of determining which pesticide to use, in the use of separate clothes for spraying, in the practice of cleaning equipment each time after use, and in the correct disposal on empty containers.

Finally, it was not possible to read the trend in three practices in the absence of baseline data. These were with regard to getting wet from spraying or leaks, washing up quickly after spraying, and washing spraying clothes after each use.

Conclusions Regarding Project Hypotheses

As indicated in Chapter 3, the findings in each country can be looked at in terms of the four hypotheses established by the Steering Committee for the project (see Chapter 1).

Hypothesis: Through communication and training we will improve practice in:

- skin protection,
- preparation of spray solution,
- washing of body and work clothes,
- spraying and application, and
- maintenance of spraying equipment.

In the test region in India, the practice of having a full body wash after spraying rose significantly and was sustained. Washing of work clothes worn while spraying was at a high level in 1996 and was sustained in 1997. The use of gloved hands for mixing pesticides showed a small improvement over 1992 but dropped sharply in 1997 after cessation of intervention. Significant improvement in the maintenance of spraying equipment was registered. Farmers' and spraymen's practice of taking precautions before breaking for food, drink, or a smoke was at a high level, and it persisted in the project area. And the use of a full-sleeved shirt and full trousers went up significantly, but after cessation of intervention these dropped somewhat, although still registering positive change.

Hypothesis: Through communication and training we will improve practice in:

- optimization of quantity of plant protection products used and of spray parameters,
- storage of plant protection products, and
- disposal of empty containers.

The practice of storing crop protection products in an acceptable manner was at a high level in both regions in India at the baseline, and that continued through 1997. The safe disposal of pesticide containers showed no improvement in either area. In fact, that situation worsened by 1997. The use of a measuring glass was at a high level in 1992, fell in 1996, and then rose again in 1997 back to the level at the baseline. The intervention programme, as was borne out by observation, ensured that a good practice more or less continued.

Hypothesis: Through communication and training we will improve practice in:

- identification of pests and beneficial insects,
- selection of suitable product,
- determination of correct dosage,
- usage of suitable equipment,
- correct timing of application, and
- proper application techniques.

Inspection of plants before taking a decision to spray and the decision regarding the pesticide to be used both improved significantly as a result of the intervention, but this change did not continue. Knowledge of beneficial insects amongst farmers and spraymen rose substantially but it dropped later (though it remained at a higher level than at the baseline). The practice of determining the correct dosage to use improved substantially and more or less continued in the project area. The commencement of spraying operations before 9:00 a.m. rose significantly among farmers and spraymen in the project region, but this positive change was not sustained.

Hypothesis: Improvement of farmers' economics will facilitate their adoption of messages on safety.

Qualitative research studies indicated that there was widespread appreciation among farmers that Novartis had put considerable effort into improving farmers' health by teaching them how to enhance their safety as well as improve their crop yields through the effective use of pesticides. This credibility of Novartis could be considered as contributing in a large measure to the positive changes noted in the adoption of safety measures and the consequent improvement in the health of the farmers and spraymen in the project area, as reported during informal interviews during a visit to the region in 1998.

Overall Conclusions

This study suggests that delineating a control region in this day and age might not be as feasible as it once was, before communications became as fast and fluid as they are today. There was clearly some spillover into the control region of the very issues that were being discussed in the project region. These could have happened through a social interaction between project and control area farmers, dealers, or their families. It could also have been the result of the better communication methods being picked up by agriculture extension workers, who might have been inspired to carry their learning into other regions.

On the other hand, if a distant control region had been selected to ensure no spillover effect, there would have been difficulty in making comparisons because of other sociocultural, climatic, or economic differences between the regions.

The other aspect of interest in the control region pertains to the quality of the learning that took place. Lessons learned in the project region showed better signs of being retained than those learned in the control region. Yet when the 1996 survey was done, it appeared that the control region had performed better than the project region. This suggests that learning by proxy spreads faster across a populace but does not stay long. Learning by proxy may run wide but shallow, in other words, while direct learning may be absorbed and go deep.

The over-riding lesson of the study in India is that in a situation of economic deprivation, economic gains will take precedence over safety or health gains among the concerned groups. Farmers will take any steps necessary to ensure a good yield. Only after they feel secure economically will they value their health. Thus, a farmer will not spend money on protective gear. If it is given free or at a subsidized rate, he might use it if it does not hamper productivity. If the gear causes sufficient discomfort to hamper work, however, he will discard it. In developing countries, then, a concerned corporation would ensure protection through devising comfortable protective gear and making these available at an affordable rate, through a subsidy.

Similarly, if a farmer can get any value at all from the sale of empty concentrate bottles, he will do so. This affords a bonus earning that can be used for those little luxuries that are usually out of reach. A concerned corporation would not try to persuade a poor farmer to give up this small bonus; it would instead organize a buyback system such that the farmer gets his small earning from selling empty containers back to the dealer, who in turn returns them to the company and receives reimbursement. This is difficult to organize, but it is the only real solution to the problem of safe disposal of empty pesticide containers.

Finally, the farmer's concern with better yields can be turned to advantage since any effort to teach a farmer about improving yield will ensure an attentive audience. This method was successfully used in this project to grab and retain farmer attention and interest. A concerned corporation would need to continue to work with farmers for their economic welfare—and to use the channel thus created to send out constant messages on the need for safety. The model used for this experiment was a good one. It would need to be turned into a module that could be easily and widely replicated at a low cost. This effort, if repeated over concentrically larger regions, would spread the message of safe and effective use of pesticides far and wide.

5

Zimbabwe

Overview of the Agricultural Sector

The land question has always been a central and controversial issue in Zimbabwe. Under the Land Tenure Act of 1969, 50% of the country was reserved for white farmers, including most of the arable land in the north and east. As a result of this policy, most of the country's communal farmers live in the lower-lying lands in the south and west, which have less rainfall and are primarily suitable for grazing. Although the act was repealed in 1979, this has made little difference to the distribution of land: about 30% is still owned by about 4,000 white commercial farmers. Acute land pressure in the communal areas has made land redistribution a pressing issue.

Despite post-independence plans to resettle 162,000 black families within three years, by 1995 only 62,000 black families had been resettled. In 1992 a Land Acquisition Act was adopted to provide for the acquisition of up to half of all white-owned land for redistribution, but the farmers affected appealed and the programme was effectively halted in 1995. In November 1997, faced with mounting domestic difficulties, the government again designated about 1,500 farms—covering about 40% of white-owned land—to be expropriated without compensation. The decision plunged the country into an economic crisis, and the redistribution programme has since been scaled down significantly, in part in response to pressure from donors that full compensation be paid for expropriated land. A more modest redistribution programme will start with farms where the government's designation has not been contested, while a donor conference is to be held to discuss the issue. As yet, however, the land question remains far from solved.

This chapter is based on material prepared by Al Imfeld, Peter Jowah, and Andreas Weder except for "Overview of the Agricultural Sector", which is excerpted with permission from Economist Intelligence Unit, Zimbabwe: Country Profile 1998–99 *(London: 1998), with minor stylistic alterations.*

Zimbabwe is generally food self-sufficient, usually exporting significant quantities of meat and maize in addition to cash crops such as tobacco, cotton, and sugar. The main harvest period for most crops is from April to September. No imports were necessary until the 1992 drought, when Zimbabwe lost 80% of its maize crop. Many commercial farms have small areas under irrigation, as do large estates on which citrus fruit, sugar, and winter wheat are grown. Further irrigation could restore wheat self-sufficiency and provide some output stability for the communal areas.

In higher rainfall areas maize yields of more than 10 metric tonnes per hectare are achieved—among the best in the world; yields in drier areas barely reach one tenth of this level. The horticultural industry has recently become one of the most dynamic exporters, and Zimbabwe is the third largest exporter of roses in the world.

Target Crop and Audience

Small-scale farmers were chosen as the target group for the project in Zimbabwe because of the widespread unease there, as elsewhere, about the use of agrochemicals given the prevalent illiteracy, lack of technical expertise, and limited extension services. Cotton was chosen as the target crop as it is the single most important crop in Zimbabwe in terms of pesticide use, especially insecticides. Use of pest control products is obligatory in the area due to the presence of a wide spectrum of pests (including bollworms, such as red and heliothis, and sucking pests, such as aphids and red spider mites) throughout the cotton-growing season.

Target Area

The lands in the test area (the Sanyati communal lands in the Kadoma District) and the control area (the Nemangwe communal lands in the Gokwe South District) are owned by the government. The principal crops in both areas are maize (the main staple food crop) and cotton (the main cash crop). The Sanyati is one of the three major communal lands in which cotton production has increased since independence in 1980, and there is considerable potential for raising the productivity of cotton there. The Sanyati communal lands lie in Natural Regions III and IV, the source of most of the cotton grown by small-scale farmers in Zimbabwe, so they experience typical conditions under which cotton is grown as a cash crop. During the project period, a shift from cash crops to food crops took place that can be explained by the drought, which affected both areas.

Sanyati is situated to the west of Harare and approximately 90 kilometres northwest of Kadoma. The total planted area of about 20,000 hectares is nearly evenly divided between cotton and maize. Some 4,000 farming families support a total population of about 29,000, and 50% of the population is literate. The

Nemangwe communal lands—the control area—are west of Kadoma, covering an area of just over 150,000 hectares, of which 55,450 are arable. Nearly 12,000 farmers there support a population of 81,000.

The liberalization of the marketing of cotton during the second half of the project saw the entrance of other competitors on the marketing scene. This put an end to the monopoly enjoyed by the Cotton Company of Zimbabwe over the years. It also resulted in farmers getting good prices for their seed cotton, which they were paid for on the spot, upon delivery.

Sanyati and Nemangwe include not just the traditional untouched Shona farmer, but also farmers who were resettled in the areas and those who moved there under their own initiative. The latter are more open to new approaches and to the European way of farming.

Farmer families have been resettled in Sanyati since the late 1950s and advised to plant cotton. The cotton scheme came later to Nemangwe, which was more heavily populated to start with and an area of more mixed farming. The two areas thus have different historical backgrounds—Sanyati is monocultural, and Nemangwe is more mixed. But common to both areas was the experience of European resettlement and the push for cotton.

Zimbabwean agricultural research discovered a difference in attitude among farmers in the two areas. Those who were resettled in Sanyati with about 5 hectares of land produced half as much per hectare as communal and smallholder farmers in Nemangwe whose farms were only 2 hectares in size on average.

Around the time the project was launched in Zimbabwe, there was a systemic lessening of agricultural support services throughout the country. It could well be that in Sanyati at least some farmers saw the intervention of the project as an instrument of the government's structural adjustment programme that began in 1991. This may explain some resistance to the project's efforts.

One problem in the research area was unfavourable weather. Sanyati, already a semiarid area, had less rain in the last 10 years than ever before. Nemangwe had the same situation, but farmers there handled it better. They had already been forced into diversifying their crops as the population in the area exploded. This experience let them adapt quicker to poor weather conditions, and Nemangwe has more intercropping than anywhere else in the country.

During the time the research was done, there was a severe drought, and the cotton crop was reduced by 65%. The maize crop was also badly affected, and the government bridged the food requirements with a drought relief programme. Many farmers in the area actually had no crop for a while. Even those who planted two crops, which is not uncommon, had nothing to show for their efforts in 1995.

More and more farmers are falling into poverty from inflation, restructuring, adaptation, weather, and soil degradation. Such a situation is not favourable for change, nor is it a fertile field for education. Thus, although the farmers in Sanyati may have enjoyed the plays they saw on pesticide issues as part of the project, and may even know the correct answers to many situations facing them, this is not a time when they want to take risks or make changes.

Baseline KAP Study

The interviews of 500 farmers in the control area and 500 farmers in the intervention area about their knowledge, attitudes, and practices (KAP) were done by Research Bureau International, under the supervision of Novartis Trading & Services. The main survey was conducted in January and February 1993, and the interviews were done in Shona or Ndebele, the local languages.

The survey found that farmers in Zimbabwe are very well informed on some issues, particularly in terms of safety, and that they follow good precautionary practices. Yet the risk to health during spraying is not a key issue, and farmers rank safety third behind effectiveness and cost when considering particular pesticides. Due to the way commerce is handled in Zimbabwe, in contrast to Mexico and India, farmers have little involvement or discretion when purchasing pesticides. They also tend to buy the supplies several months before they actually use them. Most of the farmers in Zimbabwe are influenced and informed by government extension workers.

The Concept of Health in the Target and Control Populations

It is important to be careful when trying to understand—or even worse, interpret—the meaning of health among African or Bantu people. First, a simplistic understanding will no longer do, since old and new ideas are constantly overlapping. It can no longer be said that one concept is of Bantu origin while another represents a European or developmental influence.

Second, health remains closely related to religion and philosophy. And third, it is closely linked with both ancestors and present-day society. Sickness is a warning sign that something in that relationship has been disturbed. It is thus open to interpretation and handling by the *n'anga*—the doctor of all relationships between the dead and the living, and even between nature and culture. Health is the proper responsibility of the *n'anga*.

Fourth, cotton is new and is from Europe. It therefore lies outside the domain of traditional medicine, and problems related to its cultivation belong to the European doctor. This might lead to a clash between the old and the new worlds of health.

Cotton belongs to the field of risks that can be part of health. Since health problems linked to its use are treated by European doctors, they are seen as belonging to the profane rather than the sacred. It is not hard to see, then, how the handling of pesticides is closely linked to emotions. There is no ancestral code of conduct attached to the use of pesticides. Instead, the link to the modern world leads to a strong belief that everything is possible and that accidents can be "repaired".

It is also worth noting that even among the Shona and the Ndebele, the understanding of what constitutes "health" varies widely. The Shona are by tradition agriculturalists, and they are strongly attached to their particular form of farming. For them, cotton represents the modern world. The Ndebele, on the other hand, are a fighting Zulu people, strongly attached to cattle. For them, cotton is just another tool in the fight for survival in the modern world—a fight in which you expect to suffer at least some wounds. Sores may even be interpreted as signs of openness and manly virtue. There are more Ndebele in Nemangwe than in the Sanyati area. This may explain some of the answers in our surveys that at first appeared confusing.

Intervention Programme

The main objective of the intervention programme in Zimbabwe was to develop and implement education and training methods that would promote the safe and effective use of plant protection products.

Basic Strategy

The intervention used a number of media to deliver several key messages on safety and effectiveness. The important points on safe use of pesticides dealt with the hazardous nature of the chemicals and the safeguards to follow in order to avoid accidental poisoning. Regarding effectiveness, five main messages were delivered during the programme:

- a suitable pesticide for the target pest,
- correct pesticide application timing,
- correct pesticide dosage,
- well-maintained application equipment, and
- effective pesticide coverage.

Education Through Theatre

The key messages regarding safety and effectiveness were delivered in three phases through a number of local media, but mainly through a series of plays. Plays have been used in southern Africa since 1975 as learning tools in new situations and for social development in general. They were particularly needed because more formal teaching was still associated with either apartheid or colonialism, and with paternalism and hidden persuasion. Plays have been praised as a way to create an important favourable "field of sympathy" upon which future behaviour will be based, something rooted in the community that can be shared by a whole village. Recently plays have been used as a means of sex education, especially on AIDS.

In late 1993, Cont Mhlanga—founder of the Amakhosi Production in Bulawayo, and himself a playwright with his own troupe—was asked by the project management to write a play on the safe and effective use of plant protection products that could be used as in the test area in Zimbabwe. In November 1993 and January 1994 he visited the Sanyati area, together with members of the project management team, in order to familiarize himself with the region and its farmers. He started to write a play entitled “Makanya” in February 1994.

Makanya represents an elderly farmer through whom the key messages on safe and effective use of pesticides are passed on to the audience. Following a rehearsal in mid-August 1994, the final version was agreed to and then tested at the Lozane Primary School in the Sanyati communal lands at the end of August. After the test run, a performance analysis was done with the farmers who attended the rehearsal. The play then officially opened in late November 1994 in the presence of delegates from the Ministry of Agriculture and Chiefs of local wards. Over the following three months a total of 47 performances were conducted in the Sanyati area.

Phase One (November 1994 – January 1995)

As indicated, during the play the key person Makanya communicated with his relatives and passed on messages on the safe use and application of cotton insecticides. Traditional music and occasional dances were included, which created positive interactions between the players and the audience.

At the end of the play, a quiz was held to test how well the messages had been understood. The questions were based on the main messages and on general aspects of the safe and effective use of pesticides not necessarily covered by the play or the intervention programme. Fifty questions were asked, and those who answered correctly received a T-shirt referring to “Makanya”.

The project management estimates that a third of the Sanyati population took time off to attend the play—some 2,000 men and 2,000 women. Whenever a play was staged near a school, children were also allowed to attend. Thus 7,700 boys and girls had the possibility to be involved in this phase of the intervention programme. The message delivered during such a development play will have effects when these children become mature and are responsible for farming within their communities. Thus such a play has a longer-term effect than a traditional demonstration plot.

Phase Two (November 1995 – February 1996)

The second play concentrated once more on the safe use and the application of pesticides. The programme this time was more a musical than a dramatic presentation. It was surrounded by modern African music, and a number of messages were conveyed through a dancing group. The Amakhosi Theatre Group established a lively interaction with the audience.

Again, during the musical the audience was asked the same questions as in the previous year, with the correct answer winning a T-shirt. Furthermore, a poster/calender was handed out reinforcing the key messages on the safe use and application of pesticides. At the end, all participants joined for a final dance.

The organizers as well as the actors were very pleased when the musical play reached approximately 60% of the Sanyati population.

Phase Three (September – October 1996)

Cont Mhlanga and Amakhosi Theatre Production were asked to compose a third play around the same issues. This piece was somewhat similar to European sketches or a cabaret. In between there was traditional music again. The first scene conveyed the message that the application equipment should be in good working condition. The second scene referred to the use of a suitable pesticide for the target pest, proper dosage of chemicals, and correct timing. The final scene concentrated on protective clothing, personal hygiene, and effective pesticide coverage.

As in previous plays, a quiz was given at the end. This time the focus was on reducing the use of pesticides. Again posters/calenders were given as prizes. In addition, a raffle was held for a set of protective clothing (overalls, gloves, and mouthpiece). The attendance was estimated at over 50%. Apparently the most successful item of the third phase was the cotton insect pest calculator.

Other Components of Intervention

Insect Pest Calculator

Special attention was given to scouting from the beginning of the programme. Before this, people just followed a timetable rather than observing the pests themselves. To improve effectiveness, farmers were taught to look first at the cotton plants, observe what is under the leaves, and only to spray when it is necessary. The need to spray could be defined by a cotton insect pest calculator, which was developed to reduce pesticide use substantially.

The calculator was a new tool. It was properly explained during and also after the play to people. At the end of the performance, it was handed out to farmers for their own use. This should have meant the end of spraying on a fixed schedule without any regard to an actual pest infestation. Instead, eyeball surveys of the problem should have been carried out. But people were warned that an eyeball survey alone does not indicate the best moment for spraying.

Later on, however, the project staff discovered that:

- the time after the play was too short to teach farmers the efficient use of the calculator;
- the calculator may have overshadowed some of the messages delivered in the play;
- the calculator was something so new that it was a fascinating object, almost like some modern kind of magic device;
- the calculator became a prestige object in the huts of chiefs or better-off cotton farmers, and could be seen hanging on a hut pole like a fetish; and
- people were talking about the pest calculator but not using it.

Demo Plots

In addition to the intervention programme in the form of rural plays, 10 demo plots were organized during the cotton seasons 1994/95 and 1995/96. Demo plots also have a long tradition as a teaching tool in Africa. Indeed, a half-day in each week in the school year was traditionally reserved for children to work in gardens or demo plots. (To some people, however, this meant such demonstrations were an activity for children, or were a reminder of teaching in the colonial era; some farmers were therefore less interested in the demo plots than in the plays.)

The project cotton demo fields were established at strategically well placed locations, taking the large Sanyati intervention area into consideration. Good farming knowledge of the owner was a further important point in the selection of the venue, as this person was asked to keep actual control of the cotton planting process, rainfall, growth, and so on. Two field assistants set up the trials, providing the farmers with necessary seeds, chemicals, and equipment. Based on regular scouting, the farmers were advised of the optimal spraying moment using the correct pesticide for the target crop.

In the first year, the demonstration plot yielded more cotton than a normally managed cotton field in the same area. This was interpreted as being the result of proper land preparation, careful seeding, correct use of chemical inputs, and overall timely application of pesticides used after regular scouting. It was also noted that the scouting and/or spraying decision was copied by other farmers who had regular contact with demo plot activities. Unfortunately, the demo plots established the second year, in 1995/96, were a complete failure due to the severe drought in the area.

During the time of the project, the Cotton Research Institute remained active in Zimbabwe. Since this Institute serves all cotton farmers, there may have been some exchange of information in the test and control areas, especially among the few remaining extension workers.

Posters

Three posters were produced for the project in Zimbabwe. Every household head or farmer who attended the play received a poster that depicted the following important aspects of safe and effective use of pesticides:

- protective clothing,
- reading the label,
- correct spraying,
- safe storage of pesticides,
- correct disposal of empty containers,
- personal hygiene after spraying, and
- the washing of clothes used during spraying.

In addition, the posters were given to local influential individuals and to the offices and warehouses of pesticide distributors.

The posters also pointed out:

- suitable pesticides for the target pest,
- scouting/application time,
- correct pesticides dosage,
- the importance of well-maintained application equipment, and
- effective pesticide coverage.

Results of Intervention

As in Mexico and India, the programme was evaluated by means of a qualitative study (focus group sessions) and quantitative studies, consisting of the KAP and observation surveys. The evaluation was done by Probe Market Research of Harare. The results are presented in this section.

Qualitative Evaluation

Methodology

Like the quantitative surveys, focus group sessions were conducted before and after programme implementation and in both areas in order to be able to compare developments with the programme to those without it. The endline sessions asked farmers about their understanding and appreciation of the intervention programme and the changes they perceived were occurring in their area. Unlike the quantitative surveys, the qualitative ones were not repeated one year after the conclusion of the intervention programme.

The surveys consisted of four group sessions with 10 small-scale farmers. Further in-depth interviews were held in both areas with individual farmers and local influencers.

Differences Between the Intervention and Control Areas at the Endline Survey

Based on the qualitative baseline survey, it is clear that small-scale farmers had received over many years continuous information and training by the AGRITEX (agricultural extension services), the Kadoma cotton training centre, and the Agricultural Chemical Industry Association. This training provided a good grounding in knowledge of protective gear to be used when spraying and the safe handling of pesticides.

The most important differences between the two areas at the endline can be summarized as follows:

- The claimed use of protective gear, safe use of pesticides, equipment maintenance, washing of body and working clothes after spraying, and disposal of pesticides containers showed a significantly higher incidence in the test area.
- There was more confidence regarding the choice of pesticides in the test area.
- Farmers in the control area waited until just before spraying to buy the chemicals, which could result in the product not being available or the farmer being unable to buy supplies due to a shortage of funds.

Farmers' Perception of the Changes and the Project

Farmers who benefited from the communications campaign during the intervention period indicated that they increased their knowledge in the safe use and application of pesticides. The reported use of protective gear, maintenance of spraying equipment, disposal of pesticide containers, and personal hygiene also showed an improved trend.

Overall, the intervention method used in Zimbabwe (the Makanya play) was also appreciated for its educational and entertainment value.

Quantitative Evaluation

Methodology

During the five years between the baseline and endline surveys, behaviour changed for many more or less obvious reasons. As explained in Chapter 3, the programme effect is assumed to be the difference between the changes in the test and control areas, or the "net change". A follow-up survey was conducted one year after cessation of the intervention programme, to assess the persistence of any changes detected earlier. Thus for any variable assessed all three times, there is one net change between the baseline and the endline (Net) and another change between the endline and the follow-up (Net+). Net then quantifies the proportion of farmers in the test area who have adopted a certain piece of knowledge, an attitude, or a practice as a result of the project, while Net+ quantifies the persistence of adoption. A positive Net+ indicates that adoption in the test area increased after conclusion of the project (in comparison with the control area), possibly because communication between farmers continues to spread the knowledge, attitudes, or practices in question. A Net+ value near 0 indicates that adoption stayed approximately at the level achieved when the intervention stopped, and a negative Net+ value indicates that adoption decreased after cessation of the project (always in comparison with the respective control area development).

Media Reach

At the follow-up survey, 75% of the test area farmers affirmed having received pesticide-related information or advice (compared with 44% in the control area). And 37% of test area farmers mentioned the intervention programme. A further 11% mentioned the CIBA Foundation (the predecessor to the Novartis Foundation). In addition to the play, farmers also mentioned the quiz that took place during the play, the poster/calendar on the most important issues on the safe use and application of pesticides, and the poster/calendar on beneficial insects. The demo plots were also mentioned.

Adoption of New Attitudes and Behaviour

The KAP and observation studies looked at a number of variables to assess changes in attitudes and action. Tables 5.1 to 5.5 provide overviews of the findings. As just explained, "Net" indicates the change between the baseline and endline studies, and "Net+" is for the change between the endline study and the follow-up study one year later.

Comparison of Stated and Observed Practices

Table 5.6 shows the deviation between the reported and observed practices of the endline and follow-up surveys. As the baseline observation samples were considerably smaller, those figures have not been included in the table.

Positive figures indicate practices whose reported level was higher than observed—that is, they show overclaim. Negative figures indicate practices whose reported level was lower than observed; in these cases, some farmers adopted a certain practice, but did not report it. Claimed and observed practices usually coincide within a one-digit error. Thus the two surveys corroborate one another, because although it is conceivable that both the KAP study and the observation survey (which was conducted with advance notice) are subject to reporting biases, it is unlikely that both would be subject to errors of the same magnitude.

Table 5.1. Zimbabwe: KAP Survey Results on Knowledge (percentage)

Knowledge	Test area			Control area			Test area changes		Net changes	
	B	E	F	B	E	F	E-B	F-E	Net	Net+
Face mask[a]	60	90	84	57	66	42	30	-6	21	18
Full body protection (cloth)[a]	91	95	95	91	85	85	4	0	10	0
Gloves at mixing[a]	4	67	66	2	39	28	63	-1	26	10
Gloves at spraying[a]	34	68	81	31	53	42	34	13	12	24
Gumboots/shoes[a]	72	77	91	67	61	67	5	14	11	8
Beneficials[b]	33	37	32	23	17	15	4	-5	10	-3

B = baseline E = endline F = follow-up

[a] Farmer mentions specified protection when asked about ideal protective measures to take when mixing or spraying the most toxic pesticide he has ever used.

[b] Farmer mentions that some insects are beneficial rather than being pests, on a direct question.

Table 5.2. Zimbabwe: KAP Survey Results on Attitudes (percentage)

Attitude	Test area			Control area			Test area changes		Net changes	
	B	E	F	B	E	F	E-B	F-E	Net	Net+
Health main problem[a]	11	49	62	22	50	49	38	13	10	14
Health risk exists[b]	89	80	87	81	74	86	-9	7	-2	-5
No risk, precautions taken[c]	48	73	86	54	47	35	25	13	32	25

B = baseline E = endline F = follow-up

[a] Farmer mentions health problems from spraying when asked which are the main problems associated with pesticide use.

[b] Farmer answers yes on the direct question of whether a health risk is associated with the pesticide spraying job.

[c] Farmer reports following all necessary safety precautions when asked why he believes there is no health risk associated with the spraying job, expressed as proportion of those indicating that there is no such risk.

Table 5.3. Zimbabwe: KAP Survey Results on Reported Practices (percentage)

Practice	Test area			Control area			Test area changes		Net changes	
	B	E	F	B	E	F	E-B	F-E	Net	Net+
Checked pest[a]	59	87	66	64	84	47	28	-21	8	16
Scouting scheme[b]	85	74	71	86	31	22	-11	-3	44	6
Face mask[c]	41	71	68	36	45	42	30	-3	21	0
Gloves at mixing[c]	1	37	35	1	11	6	36	-2	26	3
Gloves at spraying[c]	17	39	37	17	15	10	22	-2	24	3
Gumboots/ shoes[c]	69	78	90	62	51	62	9	12	20	1
Full body protection[c]	93	96	98	95	96	93	3	2	2	5
Cleans sprayer with soap[d]	55	64	61	61	56	50	9	-3	14	3
Container disposal[e]	76	84	80	82	70	69	8	-4	20	-3
Wash before eating[f]	86	78	93	85	80	87	-8	15	3	8

B = baseline E = endline F = follow-up

[a] When asked how he decided to use a pesticide, farmer answered "by checking pests".

[b] Farmer reports using a fixed scouting scheme, as opposed to "eyeball survey" or not inspecting the crop.

[c] Farmer reports using specified protection when asked what precautions taken while mixing or spraying last time.

[d] Farmer reports cleaning spraying equipment by pouring soap water in the tank and spraying it out.

[e] Farmer reports having burned or buried the empty pesticide container.

[f] Farmer reports taking precautions before eating, drinking, or smoking after spray work. (Further analysis shows that these precautions are washing the hands and, in few cases, the body.)

Table 5.4. Zimbabwe: KAP Survey Results on Health (percentage)

Health indicator	Test area			Control area			Test area changes		Net changes	
	B	E	F	B	E	F	E-B	F-E	Net	Net+
Headache, giddiness, nausea, ever[a]	58	20	11	56	31	18	-38	-9	-13	4
Skin problem, ever[a]	32	44	52	37	32	52	12	8	17	-12
Fever, flu, vomiting, ever[a]	22	4	24	26	1	24	-18	20	7	-3
Never health problem[b]	70	53	47	57	58	47	-17	-6	-18	5
No health problem last time[c]	67	76	71	72	43	44	9	-5	38	-6

B = baseline E = endline F = follow-up

[a] Particular problem mentioned by farmer when asked about main health problems due to pesticides, expressed as proportion of those who reported ever having such problems.

[b] Farmer answers no when asked whether he has ever suffered a health problem due to pesticides.

[c] Farmer's answer does not specify any symptom when asked whether he has suffered a health problem due to pesticide on the latest spraying occasion.

Table 5.5. Zimbabwe: Observation Survey Results on Practices (percentage)

Practice	Test area			Control area			Test area changes		Net changes	
	B	E	F	B	E	F	E-B	F-E	Net	Net+
Hat/cap/scarf /towel[a]	*	85	74	*	48	63	*	-11	*	-26
Face mask[a]	*	66	60	*	31	34	*	-6	*	-9
Special eye protection[a]	*	19	9	*	3	5	*	-10	*	-12
Overalls[a]	*	71	67	*	52	60	*	-4	*	-12
Long trousers with shirt[a]	*	77	84	*	63	65	*	7	*	5
Gumboots/leather boots[a]	*	56	50	*	29	34	*	-6	*	-11
Gloves[a]	*	43	33	*	8	8	*	-10	*	-10

B = baseline E = endline F = follow-up

* Sample size too small to be included.

[a]Observed attire while spraying.

Table 5.6. Zimbabwe: Differences between Stated and Observed Practices (percentage)

Practice	Test area		Control area	
	E	F	E	F
Hat/cap/scarf/towel	-16	4	-5	-10
Face mask	5	14	8	8
Special eye protection	-11	-2	2	-1
Overalls	3	5	6	-4
Long trousers with shirt	-10	-2	-3	1
Gumboots/ leather boots	3	6	3	6
Gloves	-4	7	4	2

E = endline F = follow-up

Across the two points in time and the two areas, the use of gloves, face masks, and overalls generally was overclaimed, while usage of head covers, long trousers, and shirts was underclaimed. A possible interpretation of this finding is that the first type of item is worn specifically for the protection from pesticides it provides, while the second is a more common type of work clothes. Thus when asked about the attire worn while spraying, farmers tend to think mostly of items worn especially on those occasions rather than their normal clothes.

In conclusion, the reported practices are a blend of actual practices and awareness of the protection provided by various gear.

Summary of Major Findings

Continued Changes Due to the Project

Behaviours and knowledge levels that changed favourably during implementation of the project and continued to increase or remained stable afterwards, in comparison both to the baseline and to corresponding developments in the control area, are the most certain to have been adopted persistently. The positive net changes allow the conclusion that these variables were favourably affected by the project. These items include the following:

- the knowledge that gumboots, cloth face masks, and gloves should be worn—the knowledge about the usefulness of overalls also falls into this category, but this item improved only a little from a very high knowledge base at the beginning of the project;
- the reported usage of gloves, gumboots/shoes, and cloth face mask—as mentioned, the reported practices reflect knowledge and awareness rather than actual behaviour;

- the reported usage of full body protection (overalls and so on), although, again, this changed only a little from a very high reported practice at the beginning of the project; and
- the perception that health risk is a major problem associated with the spraying process.

The perception that pesticide-related health risks are a problem rose by considerable proportions in both areas. Yet a small but increasing share of farmers believe that this risk is eliminated by taking all precautions.

Desired Changes That Did Not Continue

Some variables that changed in the desired direction during the project showed a reduction after cessation of the programme, assessed as net change, as test area change, or both. These behaviours or knowledge were favourably affected by the programme, but not persistently. They include the observed usage of cloth face mask, special eye protection, and gloves; the reported use of hats or caps; and the knowledge and reported usage of gumboots.

With few exceptions, the degree to which these items were forgotten or abandoned after the programme was relatively low compared with the initial adoption, suggesting that the practices will decrease only slowly.

There is no apparent correlation between the handiness of the protective items and the persistence of their adoption. Two of three farmers who mentioned some gear as being ideal but failed to use it said that they could not afford to buy the gear in question.

Changes Abandoned After Experimentation

There are no clear cases of practices that were largely abandoned after having been adopted initially. The practice of checking pests before deciding on which pesticide to buy returned to its approximate initial level after having increased, when assessed as change versus baseline. But its net changes are favourable because the practice of checking pests dropped much more in the control area.

Changes in Undesirable Direction

The perception that there is a risk associated with pesticide use was reduced, mainly because an increasing proportion of farmers think that taking all necessary precautions rules out this risk.

Other Changes

Some other variables showed favourable net changes but an unfavourable development versus the baseline, or vice versa. Others increased after but not during the programme. Changes in these variables are difficult to interpret. This is the case for scouting pests according to a fixed scheme, a practice that was almost completely abandoned in the control area according to the KAP survey, leading to favourable net changes even though the practice did not increase in the test area. Moreover, the observation and the KAP surveys yielded very different results for this variable. Evidence regarding the use of overalls was also ambiguous.

Health

The developments of the various health variables provide evidence of the difficulties involved in eliciting responses on such a complex concept as health.

The variable "headache, giddiness, nausea, ever" suggests that these most frequent symptoms were reduced, in comparison both to the baseline and to the control area. In contrast, skin problems appear to have increased in both the test and the control area. The more severe symptoms captured in "fever, flu, vomiting, ever" show a very large year-to-year variation that makes the interpretation of its development difficult. The number of health problems reported at the latest occasion of spraying also dropped, as indicated by the increase in the variable "no health problem last time", but again the control area levels show that this variable is subject to a large year-to-year variation.

It would seem that what farmers report as "health problems suffered ever" are problems suffered in the last few years rather than lifetime experiences. Theoretically, the number of farmers who have ever suffered health problems due to pesticides cannot drop by more than a certain rate implied by population dynamics. The reduction in symptoms reported to have been suffered ever clearly exceeds this rate, suggesting that farmers' memories limit the number of very remote cases reported.

Conclusions Regarding Project Hypotheses

As indicated in Chapters 3 and 4, the four hypotheses of the project can be revisited in light of the findings in Zimbabwe.

Hypothesis: Through communication and training we will improve practice in:

- skin protection,
- preparation of spray solution,
- washing of body and work clothes,
- spraying and application, and
- maintenance of spraying equipment.

Personal hygiene and skin protection were already at a high level when the project started. The project had a further positive impact on the use of gloves, proper shoe attire (gumboots and so on), and the use of full body protection. Another significant increase was reported in the regular washing of work clothes.

Regarding the mixing process, project managers recorded a much higher percentage using gloves when mixing pesticides, which was a desired result. Overall, the project had a positive impact on reported attitudes regarding personal hygiene and skin protection.

Hypothesis: Through communication and training we will improve practice in:

- optimization of quantity of plant protection products used and of spray parameters,
- storage of plant protection products, and
- disposal of empty containers.

Again, the baseline study revealed a high level of knowledge in the area of storage of chemicals and disposal of empty containers. Although there was no change in the first regard, the correct disposal of empty pesticide containers improved in the area of the intervention programme.

The cotton pest calculator was introduced in an attempt to improve the use of the optimum quantity of pesticide. The result was not noticeable, however, as this tool needs a more comprehensive introductory programme, including field training working with heavy pest infestation. Such training was not possible during the project period, and the managers realized that the training given to farmers during the play did not result in correct use of the calculator under actual field conditions.

Hypothesis: Through communication and training we will improve practice in:

- identification of pests and beneficial insects,
- selection of suitable product,
- determination of correct dosage,
- usage of suitable equipment,
- correct timing of application, and
- proper application techniques.

It is not easy to summarize the improvement in this area due to the project. Farmers received chemicals in the past through the credit scheme. The impact of the liberalized market on the use of suitable products remains to be seen. Knowledge of beneficial insects remained at a moderate level. But only half the farmers reported that they keep up with regular maintenance of equipment.

Hypothesis: Improvement of farmers' economics will facilitate their adoption of messages on safety.

The continuous changes in the climate from year to year and the severe drought in 1995/96, which meant the loss of food as well as cash crops, had a considerable negative impact on the farmers' economic situations. In general, available funds are used first for basic needs (daily operating expenses, for example, or schooling for children) before investments are made in the field of safety.

It is too early to tell if the liberalized cotton market will improve the financial situation of cotton farmers enough to allow them to purchase adequate safety gear and spraying equipment.

Overall Conclusions

The various practices promoted in the rural theatre programme during the project improved farmers' knowledge about safety and, to a lesser extent, their practices. It is important to remember that Zimbabwean farmers already had a high level of knowledge about safety when the research programme began, thanks to the activities of agricultural extension workers.

During the project, project staff noticed that farmers tended to answer interviewers' questions in ways that would please the interviewer. In other words, they passed on knowledge of correct procedures as actual practice. A further overclaim of correct practices was noted when advance notice was given that farmers would be observed by the project staff.

Both the qualitative and the quantitative evaluations showed that the project achieved a fair number of desired changes regarding knowledge, attitudes, and—to a lesser extent—practices. A constant application of the knowledge gained will have a favourable effect on farmers' health and their economic situations, due to higher crop yields or reduced application of pesticides.

The positive impact of the project in Zimbabwe will be even more discernible in the future, as the changes that have been noted are mostly sustainable and are likely to continue after conclusion of the project.

6

Conclusions

In the next 30 years the world will have to feed at least 2 billion more people. This will have to be done with less arable land, fewer nonrenewable resources such as phosphorus and potassium, less water, and a poorer quality of water, as well as with fewer and older people working in the agricultural sector.

Most experts today believe that future increases in food production in developing countries will have to come mainly from yield increases rather than from expansion of the land devoted to agriculture. Yet there are signs that give rise to concern: in major areas the yield increases of principal food crops are at best stagnating, and they are declining in areas where intensification is most developed.

One obvious conclusion is that losses due to pests, weeds, or fungi have to be minimized in order to make sure that a maximum of the potential harvest is turned into actual harvest. This means that pesticides, for the time being, will not only have their place in the agriculture of developing countries, but will also remain important. Farmers in all developing regions choose and demand synthetic pesticides as one technology for limiting their crop losses. They will continue to do so for the foreseeable future. When pesticide prices rose sharply after the Mexican peso devaluation, producers did not shift to nonchemical approaches but instead substituted cheaper, more hazardous pesticides for the brands that they had used before the devaluation. Thus regardless of external circumstances, one continuing issue will be to balance the benefits of using pesticides with the costs—including the harmful effects on the health of users, a subject addressed by this project.

The project on safe and effective use of pesticides wanted to clarify:

- what factors hinder the safe and effective use of pesticides in developing countries;
- what sort of groundwork can or must manufacturers, in collaboration with other institutions, lay to eliminate these factors; and
- what, in a given social cultural context, are the communication methods best suited to the farmer community.

Research across three continents and thereby in different cultures was done in order to evaluate the importance of differences in sociocultural situations. The safety and effectiveness problems, however, seemed to be very similar. Even in comparisons of food crops (maize in Mexico) versus cash crops (cotton in India and Zimbabwe) and of spraymen versus farmers (in India), there were important similarities. The great differences lay in attitudes due to differences in values and culture and resulting differences in symbols (the use of an apron, for example, and taboos).

It turned out to be very difficult to assess or measure the impact of isolated intervention in a dynamic social environment where life in general changes constantly through improved communication and where competitive companies and agricultural extension services are active. The programme over such an extended period of time in such diverse cultures led to a learning process that raised additional questions and issues to be analysed, and thereby some issues that later on were found to be of importance were not covered by the initial baseline study.

According to the experience of empirical social research, we know that people have a tendency to give answers they think are desirable from the point of the interviewer. This is why we did controls through the observation of actual behaviour. We found considerable differences between the claimed behaviour (interviews) and the actual behaviour (observations). When the observation data were collected, it was found that often when farmers were given advance notice (as was necessary), they tended to behave better than they would have normally. This means that the difference between the claimed and actual behaviours could be even larger. In many fields we could not refer to pre-existing research, and thereby had to take some risks. Despite the problems that were encountered because of the pioneering nature of the work, we would encourage future researchers to undertake similar projects that cut across cultures.

Our initial surveys made it clear that most of the farmers in the target areas in the three countries under review were aware of pesticide-related health risks. More than three quarters of the producers reported that they had not suffered any ill effects from using pesticides. At the same time, though, anywhere from 14% of the pesticide users in India to 25% in Mexico reported health problems that they attributed to using pesticides. In India the complaints were about skin problems, headaches, and giddiness, while in Zimbabwe they were about headaches, vomiting, and nausea. The absence of medical records made it impossible to assess the validity or seriousness of these complaints, however.

The surveys also indicated that prior to any interventions, the level of knowledge about risk-reducing precautions varied by region and by topic. Broadly speaking, the farmers in all regions seemed to be aware of the importance of personal hygiene (such as washing hands after spraying) and storing their pesticides carefully, as well as the need to keep their sprayers clean. At the same time, though, there were differences in awareness of the need to wear masks and footwear while spraying, or clothing to provide adequate body coverage. In general, though, whatever knowledge farmers had was not necessarily translated into practice at the field level. This applied especially to such practices as wearing additional clothing (aprons, special boots, and gloves).

Although the nature of the data makes it difficult to be categorical, it appears that the overall level of knowledge about what should be done to reduce health risks was relatively low and the proportion of farmers who adopted appropriate practices was even lower. The gap between the reported levels of knowledge and actual practices makes it clear that knowledge alone was not necessarily the most important factor stopping farmers from taking the essential steps to reduce their health risks. The economics of using pesticides appeared to be more important to them than the possible health impacts.

The interventions that were intended to increase producers' knowledge and have them adopt good practices consisted primarily of informing farmers, through various means, about the importance of taking precautions when using pesticides. The translation of higher knowledge and improved awareness into a desirable change of practice depends on numerous and complex variables such as affordability, climatic condition, cultural factors, simplicity, felt need, comfort/discomfort, and availability as well as general political and economical conditions.

Overall, the interventions—which ranged from intensive training of farmers in India to the use of folk theatre in Zimbabwe—did have a positive impact. We learned some important lessons during the course of the project. The most fundamental one was that messages need to focus on practical, basic, ready-to-use, but effective recommendations. Suggesting the use of impractical or expensive items or habits can dilute the overall message about safety precautions. In addition, the mix of communications media used in each country kept being refined during the project, reinforcing the point that social marketing campaigns need to be tailored to specific locations.

In each case the media chosen were those that were traditional to the area—films in India, radio in Mexico, and plays in Zimbabwe. The project thus relied on entertainment to educate, rather than books and more formal schooling. Different levels of learning also proved effective: learning by listening and watching an entertaining sketch or play, learning by observing a demonstration plot, and learning by doing, as in the case of the pest calculator the project tried to introduce in Zimbabwe.

Special efforts were made to include in the communications campaign children of the farmers who use pesticides, which was also innovative. In Mexico, a playbook and cartoons were used, while in Zimbabwe children were encouraged to attend the plays. A series of programmes for rural schoolchildren was introduced in India. This not only helped spread the message to the parents, it also served to prepare tomorrow's farmers to use pesticides safely and effectively. The project's use of these communications media could be called ground-breaking.

Any comparison of safety adoption patterns in the three countries studied should ideally be conducted in terms of changes in health. Unfortunately, health was difficult to assess, as discussed in Chapter 2; indicators have to be used instead. In addition, ideally we would have measured improvements in effectiveness, but this involves cost-benefit analysis (CBA). As described in Chapter 2, the CBA results we developed were not reliable enough to make a valid and practical contribution to the interpretation of the project's results.

For knowledge and practices, the average adoption value of the variables presented in Chapters 3, 4, and 5 was calculated. (See Figure 6.1.) Each indicator is depicted for the three points in time (baseline, endline, and follow-up survey) and the two areas (test and control). The net changes or differences between the developments in the two areas are also shown. Attitudes are omitted because the concepts measured by the different attitude variables seemed too different to allow calculation of an "attitude average".

Note that the sets of variables used in each country partly diverge and partly overlap. (See Table 6.1.) This is because the forms but also the contents of the communication programmes were adapted locally, so the most meaningful indicators for one country are not necessarily the best variables for another. It also implies that the absolute value of the averages cannot be compared directly between countries. The qualitative developments over time and the net changes can be compared, however.

In Zimbabwe, adoption patterns are pretty much as we would have expected them at the start of the project. The knowledge level increased in the test area between the baseline and the endline and between the endline and follow-up. The practices followed a similar pattern. The control area levels were quite stable. The test area improvements therefore translated into favourable net changes. Incremental adoption in the test area was most likely due to the project, and it persisted after the end of the programme.

India differs from Zimbabwe in two respects. First, the control area levels were less stable and tended to follow the test area levels, especially as far as knowledge is concerned. This suggests some spillover—a plausible conclusion because there was no geographical barrier between the two areas. Second, the adoption of some practices returned to the initial levels after cessation of the programme. The average net change is accordingly positive at the endline but disappears by the time of the follow-up. Knowledge levels also improved. The net change persisted, but only because control area levels dropped as fast as those in the test area. Overall, the positive impact was only temporary. This might be partly due to the subsidies provided for protective gear, which were given initially but later eliminated. (The subsidies were unique to the Indian programme.)

Figure 6.1. Adoption Patterns of Knowledge and Reported Practices in Study Areas

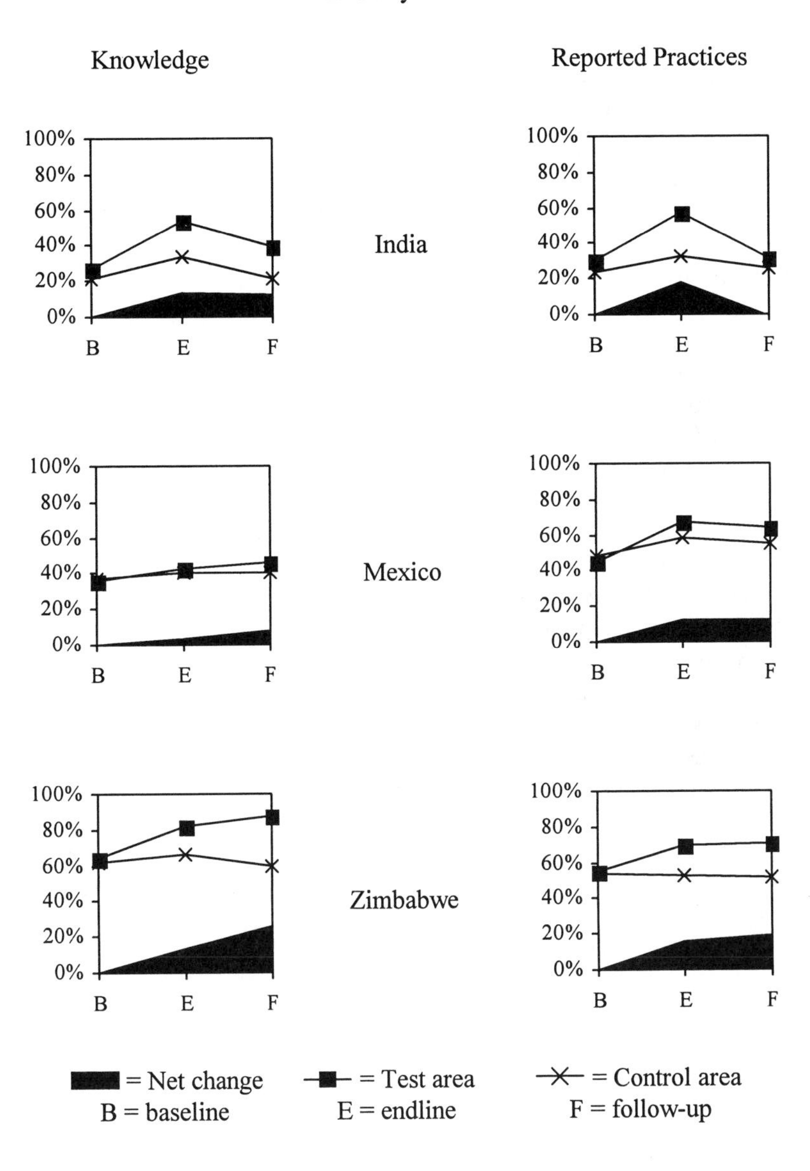

Table 6.1. Variables Used for Comparison of Safety Adoption Patterns

	India	Mexico	Zimbabwe
Knowledge	Face mask Overalls Gloves at mixing Full trousers Full shirt	Pesticides not harmless Toxicity indicated by stripe Boots Gloves at mixing Apron	Face mask Full body protection Gloves at spraying Gumboots / shoes
Reported practices	Washed before eating Face mask Overalls Full trousers Full shirt Cleans sprayer with water	Boots Wash hands Dosifier No chili can Cleans sprayer No leaks Out of children's reach	Face mask Gloves at spraying Gumboots Full body protection Cleans sprayer with soap

See Tables 3.11, 3.13, 4.4, 4.6, 5.1, and 5.3 for a description of the variables and data.

In Mexico, the average knowledge level showed little variation between areas and years. Nevertheless, practices improved in both areas. This supports the conclusion that motivation was lacking rather than knowledge—a finding that was clear after the first intervention season and that led to the inclusion of more motivational elements into the communication programme. The improvement of practices was more pronounced in the test area, producing a favourable net change. Net adoption persisted after conclusion of the programme, despite a small reduction in the absolute level of the indicator. Control area practices improved temporarily. As noted in Chapter 3, this was probably the result of a marketing campaign of another agrochemical manufacturer that took place in Cintalapa (the control) but not in the Villaflores area.

With due care for the limitations of a quantitative comparison between countries, it can be said that safer practices were adopted in all countries. The average incremental adoption of five to seven key practices was surprisingly similar between countries, in the range between 10 and 20 percentage points, while knowledge adoption patterns varied between countries. Nevertheless, it seems clear that there is an interaction between admitting the problem and acquiring knowledge about solving it: a willingness to admit a problem facilitates learning how to address it, and learning about a solution makes it easier to see and admit the problem.

Thus it appears that well-designed programmes introduced to educate, train, and inform farmers about simple procedures to reduce the health risks of using pesticides can indeed have a positive impact. This seemed to apply with special force where the procedures to be adopted were simple, inexpensive, and did not involve such items as wearing heavy clothing in a tropical environment. Wearing long-sleeved shirts and long trousers seemed to be adequate substitutes for aprons and the like, although they may not be sufficient for professional spraymen (as in India).

Despite the increase in the number of farmers adopting improved practices, a large number still did not do so even though they were aware of the health risks. There are at least three possible explanations for the persistence of the phenomenon. The first is that the "early adopters" or those most susceptible to change may have adopted some of the changes, leaving the laggards yet to be convinced. (This will necessarily take more time.) The second is that the producers in the regions are very poor so that they are risk-averse and cannot take chances about the possible consequences of change, especially if the changes involve financial outlays. For example, the purchase of protective gear fell sharply after the subsidy was removed in India; conversely, though, few farmers in one area used gloves (a recommended safety precaution) even though they were made available at no cost to the users by a major pesticide company.

A third reason for farmers not taking precautions even though they were aware of the importance of doing so was that "external forces" overwhelmed rational behaviour. Thus, the drought in Zimbabwe and the sharp devaluation in Mexico disrupted all farming activities, including those related to the use of pesticides. These may well have been short-term disruptions, however, so their long-term effects are still open to question.

Besides changed habits, better quality and better maintenance of spraying equipment is one of the most important safety measures that was adopted.

One somewhat sobering result of the project is that social marketing campaigns have to be done on a sustained basis, as farmers tend to fall back into undesirable attitudes and habits after some time. Change cannot be maintained with time-restricted interventions only. There is a pronounced need for ongoing intervention to ensure persistent change.

But this is not necessarily bad news for the companies involved, because intervention programmes that are credible and problem-orientated improve confidence in the agricultural company behind them. They help the reputation of the company and thereby enhance its image, which may favourably influence the attractiveness of the company's products in the marketplace. This at least partly compensates for the costs of the programme. On the other hand, given the heterogeneity of pesticide manufacturers and taking into account that there are "rich" companies as well as "poor" ones, it is not likely that all manufacturers will do a comparable job in changing the state of affairs.

The problem arising from that is that isolated endeavours from single manufacturers can have only a very limited effect. Complex problems such as unsafe use of pesticides in developing countries demand not only actions from transnational corporations but an industry-wide approach, otherwise only a small segment of the market and hence a small number of farmers benefit. Only a consistent multistakeholder approach involving national as well as international corporations, trade distribution channels, regulatory authorities, nongovernmental organizations, farmers' organizations, and extension units can make a difference. As the sustained funding of a project such as this is too expensive for any individual company, these programmes should be supported by either an industry association or an agricultural extension agency, or by public-private partnerships.

It needs to be said that the private-industry efforts suggested can only lead to meaningful results if the economic conditions of the large number of poor smallholders improve. This calls for an economic and agricultural policy that provides at least minimum security and stability for risk-averse smallholders. A prerequisite for this, however, is agricultural policies that favour this group of farmers. International organizations such as the World Trade Organization, World Bank, and International Monetary Fund therefore need to support governments in their efforts to design and implement policies that are favourable to smallholders. A suitable environment for a safer and more effective use of pesticides could be created by making this group the target of agricultural policies and support programmes.

An interesting detailed result of the study is that a highly technical and expensive approach is not needed to improve safety. Such an approach is neither feasible nor sustainable. The good news is that a small number of simple changes make a big difference and this, according to our experience, is feasible.

Another important lesson that we learned is that safety messages in isolation are not likely to have a big impact, but when delivered in conjunction with an effectiveness message, the safety point is much more appealing to the farmer. (The farmer, however, analyses and assesses the package, and discards components according to a number of factors). Messages that do not exclusively focus on the farmers but also involve his social environment (the family, peers, and so on) are more successful as they trigger additional pressure and heightened awareness about the issue. An attractive intervention programme that seems to be in the overall interest of the farmer will create a ripple effect over and above the intervention group or area, leading to a broader impact on society.

In conclusion, the project draws attention to the fact that if farmers were to take a series of relatively simple steps, they could reduce their exposure to pesticide-related health risks. At present, many if not most producers give low priority to "safety", and many have not adopted the necessary precautions to reduce health risks. Some procedures may well be made more acceptable to low-income farmers—for example, by developing and subsidizing the sale of both cheap and comfortable clothing that can provide adequate dermal protection. In the main, though, it appears that there are few if any easy ways to promote change among large numbers of poor smallholders.

There will have to be a continued reliance on sustained efforts such as some of those incorporated in this project. But all available experience indicates that there are limits to the extent to which changes will be adopted within a generation. Even the best and most sustained efforts run into the paradoxical situation that not everyone who can adopt relatively simple modifications in behaviour will actually do so, even when it is shown that the changes are in the person's long-term best interests. Given that, any pesticide manufacturer that cannot guarantee the safe handling and use of its toxicity class 1A and 1B products should withdraw those products from the market. At the same time, since in all likelihood pesticides will continue to be an essential crop protection tool in the years ahead, there is a continued need to get farmers to adopt the most important risk-reducing procedures.

Appendix

Endline Questionnaire

This appendix contains the questionnaire used in India in 1997 in the endline study. There were minor variations on the forms used in the three countries due to cultural differences, but otherwise they were the same.

QUANTITATIVE QUESTIONNAIRE - FARMERS

1-10

Contact number						11-15
Name of respondent						
Address						
Village name						

16-17

Region			
Control	1	ZoneA	ZoneB
ZoneC			
Intervention	2		

18-19

	Yes	No
Small farmer (less than 5 acres)	1	2
Grows cotton	1	2
Uses chemical pesticides	1	2

20

Gender	M	F

21

Name of interviewer ______________________________ Date
Signature ______________________________
22-27

Name of supervisor ______________________________ Date
Signature ______________________________
28-33

Greetings. We have come from the Social and Rural Research Institute and we carry out studies on a range of issues related to farming. Today, we wish to talk to you about some of your own practices and problems in the area of farming. May we have a few minutes of your time ? (*This is the time for the contact phase: if respondent qualifies for the main interview, you will have to ask for more time at that stage.*)

1a. To begin with, referring to this field that you are working on. What are the crops that you are growing this season? And what did you grow in the same season last year? And what did you grow in the same season in the year before that and in the year before that, for the same season?

	This year (1997)	Last year (1996)	In 1995	In 1994
Cotton	1	1	1	1
Bengal gram	2	2	2	2
Vegetables	3	3	3	3
Maize	4	4	4	4
Others (specify)				
	34-43	44-53	54-63	64-73

IF COTTON NOT MENTIONED AT ALL GO TO FILTER GRID.

1b. Currently on how much area do you grow cotton ?
______________________acres hectares. 74-77

TO QUALIFY, THE GROWER MUST HAVE SOWN NO MORE THAN FIVE ACRES OF COTTON.

1c. Do you use chemical pesticide on your crop ?

Yes 1
No 2 78

1d. Who mainly does the spraying?

Self 1
Others 2 *Note response to next question with care*
79

2. Are you involved in the decisions regarding the purchase of pesticide? And have you been involved in the actual use of pesticides for even one spraying round this season?

Involved in purchase but not in use	1	**TERMINATE**
Involved in use but not in purchase	2	
Involved in both purchase and use	3	
Neither in purchase nor in use	4	**TERMINATE**
		80

3. When was the most recent occasion on which pesticides were sprayed on your farm?

_____________________________days ago. 81-82

IF OVER 30 DAYS AGO, TERMINATE.

4. Who is the person who mainly mixes and sprays pesticides ?

	Mixes		Sprays	
Self	1		1	
Family member	2		2	
Paid person	3		3	
Others	4	83	4	84

FILTER GRID

A. Has he grown cotton			Yes	No
	This year		1	2
	Last year		1	2
	The year before		1	2
	The year before that		1	2
				85-88
B. This land on which the crop is being grown.	Belongs to him		1	
	Belongs to somebody else but has been leased by him		2	
	Belongs to somebody else and he is paid to farm it		3	
	None of the above but (specify)		4	
			89	
C. Is he involved in use of pesticides	Yes	No		
	1	2	90	

TO QUALIFY, THE RESPONDENT MUST
A: Have at least two "yes" answers B: Have worked on the farm, whether his own or somebody else's C: Must have a "yes" answer

MAIN QUESTIONNAIRE

Contact number						101-105
Name of respondent						
Village name						

IMPORTANT: Transfer number from the contact questionnaire

1a. As a person involved with farming, can you tell me what you consider to be your main problems with regard to farming? (*Rank in order of mention*) *DO NOT PROMPT*

	Rank	Not mentioned
Water availability/rain	________	1
Labour	________	2
Pest infestation	________	3
Low price of produce	________	4
High cost of inputs	________	5
Health problems due to cost of pesticides	________	6
	______	106-111

1b. *If pest infestation not spontaneously mentioned, ask*:
Do you consider pest infestation to be a problem vis-a-vis other problems?
DO NOT PROMPT

Code that response which comes closest to the actual answer given. If none are suitable, write verbatim answer under "others".

	Single code only
Yes, it really is a major problem	1
Yes, it is a problem but not a major one	2
Yes, but it is a relatively minor problem	3
No, it is not a problem	4

Don't know/cannot say (DK/CS) 5
Other responses ______________________________
____112

1c. Which pests i.e., weeds, insects or disease, if any, are troubling your crop at present? *Multiple codes OK*
1d. *For each one coded under col.1c., ask*: What is the name of the weed/insect/disease that is troubling your crop nowadays?

	Column 1c At present	Name 1	Name 2	Name 3	
Weed	1	________	_____	__________	117-122
Insects	2	________	_____	__________	123-128
Disease	3	________	_____	__________	129-134
None of these	4	________	_____	__________	
DK/CS	5	________	_____	__________	
Depends	6	________	_____	__________	
Others		________	_____	__________	
	113				

SHOW CARD "A"

2. You said you are growing cotton in your farm just now. Please look at these illustrations and tell me which stage of growth your cotton crop is in just now.

SINGLE CODE

Growth Stage	1	2	3	4	5	6	DK/CS
						135	

3a. At what stage did you first use pesticide?

Growth stage number ____________________________ 136

SHOW ALBUM OF PESTICIDE PHOTOGRAPHS.
3b. Which pesticide did you use on this last spraying occasion?

Name of pesticide:______________________________________ 137-144
Code:______________________________________

4a. How did you decide that you needed to use pesticides? *Code response: now ask:*
4b. How did you decide which pesticide to use? *Multiple responses okay in col. 4a and 4b.*

	4a		4b
Inspected the plant and saw pests	1*	Identified the pest	1
Inspected the plant and saw disease	2*	Always use this pesticide	2
Inspected the plant and saw weeds	3*	Asked for advice after identifying the pest	3
Saw symptoms of pest attach	4*	After looking at what others were spraying	4
Because pesticide used when plant has grown to this size	5	After looking at what was available	5
Because others were using	6	After looking at what was less expensive	6
Because it was time to use it	7	DK/CS	7
Was advised to do so	8	Others	8
Others	9		
	145-154		

15-16

** If 1-4 coded:*
Mention how plant was inspected:
Fixed scouting scheme 1
Eyeball survey 2 168

SHOW ALBUM OF PEST ILLUSTRATIONS
5a. Which weed/insects/disease have attacked your crop so far?
5b. *If more than one, ask*: Which is your crop currently most widely infested with?
Code from illustrations or write names

Weeds/insects/diseases mentioned:____________________________________

____

____ 166-173

Main weed/insects/diseases mentioned: (single code)
____174-175

6a. What is the type of spraying equipment that you are using?

Type of sprayer:________________________________ 176-177
Type of nozzle:_________________________________ 178-179

6b. What is the capacity of the spray tank that you are using?

____ litres 180-182

6c. How good or bad is the condition of this equipment? *DO NOT PROMPT*

Works perfectly	1
Has a defective nozzle	2
Leaking tank	3
Leaks from __________________	4
Pump not effective	5
Too noisy	6
Too heavy	7
DK/CS	8
Others	9
__________________	183-
__________________	192

Okay, I now want to ask some questions regarding your view of pesticides in general

7a. What are the sources from which you mainly get information with regard to farming? (*multiple codes OK*)
7b. And of these, which do you prefer as the best source? (*Single code*)

	7a	7b
Agricultural extension officer	1	1
Dealer	2	2
Other farmers like me	3	3
Other bigger farmers	4	4
Older farmers	5	5
None/nobody	6	6
DK/CS	7	7
Others (specify)	8	8
	201-210	211-212

8. There are three main attributes that people look for in pesticides. These are: (*Read out list given below. Rotate order of mention.*) Of these, which do you consider to be the most important attribute of a pesticide? And which do you consider to be the second most important attribute?

	Rank			
Cost	1	2	3	213
Effectiveness	1	2	3	214
Safe for humans	1	2	3	215

9. What are the sources or who are the persons normally consulted before you purchase pesticides? (*Code as mentioned*: *DO NOT PROMPT*: *Multiple codes OK*)

Company representative	01
Dealer	02
Other farmers (small)	03
Other farmers (large)	04
Agricultural extension worker	05
Radio	06
TV	07
Label on the pack	08
Nobody/nothing	09
DK/CS	10

Others (specify)______________________________

216-225

10. How long before use do you normally buy the pesticide? (*single code*)

Less than 24 hours	1
1-3 days	2
4-7 days	3
8-15 days	4
16+ days	5
DK/CS	6

Others (specify)______________________________

____

____ 226-227

11. How do you decide what quantity of pesticide to buy at one time?

Plant size	01
Extent of problem	02
Enough for one day's spraying	03

Depends on cash available 04
Depends on pesticide available 05

Enough for entire season 06
Enough for a fixed period 07 Specify:______________

____
238-239 Unit code 240
Enough for one round of my field 08
DK/CS 09
Others (specify)__________________
__________________ 228-237

12a. The last time you went to buy pesticides, what are the issues that you inquired about?
12b. What details mentioned on the pack did you take into consideration?

	12a	12b
Nothing	01	01
Dosage to be administered	02	02
The particular pest infestation	03	03
Manufacturing company	04	04
Price	05	05
Antidote	06	06
Expiry date	07	07
Toxicity write up	08	08
Toxicity symbols/signs	09	09
DK/CS	10	10
Others (specify)______	11	11
__________________	241-250	251-260

If "7" or "8" i.e. Expiry data or Toxicity write up are not coded in Q.12a or Q.12b ask,
13a. Is the expiry data of the pesticide mentioned on the pack?
13b. Is the level of toxicity or poisonousness indicated on the pack?

	Expiry date		Level of toxicity	
Yes	1		1	
No	2		2	
DK/CS	3	262	3	263

If 1 coded under level of toxicity or Toxicity symbols/signs coded in Q.12a or Q.12b (coded "8" or "9" in Q.12b), ask

13c. How is this level indicated?

(*Note as mentioned: DO NOT PROMPT: Multiple codes OK*)

It is written on the label	1	
There is a symbol (mark, design) on the label	2	
There are different colours on the label	3	
DK/CS	4	264-265

If 3 coded in Q.13c, ask:

13d. Can you tell me what colour indicates the most poisonous pesticide and what colour indicates the least poisonous pesticide?

	Colour	
Most poisonous (toxic) pesticide	______________	266
Least poisonous (toxic) pesticide	______________	267

14. When you buy pesticides, how do you store the unopened bottles?

(*Code as mentioned, do not prompt)*

Not stored, used immediately	01
Stored in a shed	02
Stored on a shelf	03
Stored in a place that is out of reach	04
Stored in a place that is out of normal living area	05
No fixed place, kept anywhere	06
DK/CS	07
Others (specify)______________	
	268-277

Now I want to talk in detail of this last occasion on which you sprayed pesticides OR on which pesticides were sprayed on your farm.

15a. On this last occasion that you sprayed _____days ago (*note from Q3.*), which insect/weed/disease did you spray against?

Name	Code Number
Insect_____________________	278-281
Weed_____________________	282-285
Disease _____________________	286-289

15b. Which pesticide did you use on this last spraying occasion?

Against	Name	Code Number
Insect	_____________________	290-295
Weed	_____________________	301-306
Disease	_____________________	307-312

15c. What volume of (each) pesticide did you use that day?

Name	Volume
___________________________________	ml/litre
___________________________________	ml/litre
___________________________________	ml/litre

331-333

15d. What acreage did you spray? _______________________acres 331-333

15e. How many tanks did you spray in all that day?
_____________________tanks 334-336

16a. On that day, did you measure the pesticide?

No, used judgement only	1	
Used entire contents from the pesticide container	2	
Yes, used after measuring	3	337

If "3 *coded in Q16a, ask:*
16b. How did you measure the pesticide?

Fertilizer cup	1	
Ladle	2	
Bottle lid	3	
Measuring glass	4	
Others (specify)	5	338-339

16c. On this last occasion when you sprayed:

Yes No

IF YES:

Did you make the entire required mixture for the day at one time? How was the total volume decided?

1 2 341-346

340

Did you make a fresh mixture as required for each tank?

How many times did you make the mixture in all? 1

2 347 348-353

Other replies

Note details:

354-361

16d. What factors did you take into consideration when you decided on the quantity of pesticide and water to be mixed?

	Multicodes okay
Land area	01
Recommended dosage	02
Crop growth stage	03
Crop condition	04
Degree of infestation	05
Pesticide concentrate	06
Advice from other farmers, friends	07
Advice from agri extension worker	08
Learnt from meetings, video, play last year	09
DK/CS	10
Others (specify)________________________	362-371

Ask Q.16d1-Q.16d2 if coded "9" in the Q.16d.

Q.16d1. When did you attend this meetings (or) when did you watch the plays or videos on the mixing of pesticides? 372-375

Q.16d2. What details could you recall from the meeting(s) you attended or the videos and plays which you watched on the above said subject?

376-381

Ask all

16e. From where did you get the information about the preparation of spray mixtures?

Learnt from pesticide label	01	
Learnt from experience, judgement	02	
Taught by ______________	03	388-389
Others (specify)	04	

382-387

17a. This recent time when you used pesticides, who did the mixing of the pesticide solution?

Respondent alone	1	
Respondent + helper /assistant	2	
Helper/assistant/sprayman alone	3	
DK/CS	4	
Others (specify)______________________		390-391

17b. How did you (or the helper) stir/mix the pesticide?

Used bare hands, without stick	1
Used gloved hands, without stick	2
Used bare hands, with stick	3
Used gloved hands, with stick	4
Used knapsack handle	5
Shaking the knapsack/tank	6
Filling the tank with water and concentrate alternatively	7

392-393

18a. Different farmers have mentioned different preparatory activities prior to spraying. Can you tell me with reference to the last time that you sprayed, did you change into different clothes for the spraying work?

401

Yes	1
No	2
DK	3

If "1" coded, ask:

18b. Can you describe these items of special clothes?

Any old clothes that are to be discarded	1
Clothes specially kept aside for spraying	2
Others (describe) ____________	402-409

19a. Thinking back to that day, what kind of day was it, i.e., was it windy, sunny, rainy, etc. ?

Windy	1	Sunny	1	Very hot	1
Still	2	Cloudy	2	Hot	2
	410		411	Pleasant	3
				412	

19b. On that day, at what time approximately did you begin spraying?

19c. And at what time approximately did you stop spraying?

Delete one

Starting time	am	p.m.	DK/CS	413-414
Stopping time	am	p.m.	DK/CS	415-416

19d. And did you take any breaks in this period, other than for refilling the tank?

Yes	1
No	2
DK	3

417

If yes

19e. How many breaks did you take in all?

_________________________________ breaks

19f. What were the reasons for these breaks? And approximately how long did each break last?

Break	Reason					Duration (in minutes)
No.	Smoke	Food	Drink	Rest	Other	
1						
2						
3						
4						
5						
6						

418-460

If smoke, food or drink coded under reasons, ask:
19g. Did you take any precautions before food, smoke or drink?

Yes	1	
No	2	461

If yes,
19h. What were these precautions?

____

____

____

462-467

Code below after writing verbatim answer

Mentions washing		1	Code below
Does not mention washing 2	468		

Washed	Water only	Water + soap	
Hands	1	2	469-472
Feet	1	2	
Face	1	2	
Mouth	1	x	

20a. During this recent occasion when you sprayed, what measures did you take to protect yourself? Any other? *Code below in Col.20a. as stated. DO NOT PROMPT.*

Protection used:	20a
Wore full shirt	01
Wore full trousers	02
Wore cloth mask on nose and mouth	03
Wore plastic face visor	04
Wore hat/cap	05
Wore gloves for mixing	06
Wore cloth apron - front only	07
Wore cloth apron - front and back	08
Wore plastic apron - front only	09
Wore plastic apron - front and back	10
Wore plastic shoulder cover	11
Sprayed along wind direction	12
Did not smoke while spraying	13
Did not spray continuously	14
DK/CS	15
Others (specify) ________________	473-492

20b. Did you use a hat or cap?

Hat	1	
Cap	2	
Nothing	3	493-494

20c. Did you use anything to protect your nose, mouth, eyes?

Handkerchief	1
Face mask	2
Goggles	3
Mask (nose / mouth)	4
Nothing	5
495-497	

20d. If you wore a shirt that day, was it full sleeved or short sleeved?

Did not wear a shirt	1
No sleeves	2
Short sleeved	3
Full sleeved	4
498	

20e. Did you wear something over the clothes (*as answered in Q20d.*)

Nylon/plastic cover - front		01
Nylon/plastic cover - front & back		02
Others (specify)		03
Nothing		04
	501-507	

20f. What did you wear on the lower torso?

Shorts		1
Full trousers		2
Lungi - full length		3
Lungi - half		4
Others		5
	508-511	

20g. What if anything, did you wear on your feet that day?

Shoes (canvas, tennis)	1
Rubber boots	2
Sandals	3
Chappals	4
Nothing	5

512-513

21a. In the course of this recent spraying operation, did your clothes get wet due to any leaks or spills or the spray?

21b. And did <u>any part of your body</u> get wet due to leaks, spills or the spray?

Got wet:		Due to:	
Clothes	1	1	1
Body: full	2	2	2
Face	3	3	3
Hands	4	4	4
Arms	5	5	5
Feet	6	6	6
Legs	7	7	7
Chest	8	8	8
	514-518	519-523	524-528

22a. How did you clean the spraying equipment?

Wiped with a wet cloth	01
Poured water in tank and spray out	02
Poured soap water in tank and spray out	03
Did not clean at all	04
Not involved, not aware	05

Others (specify)______________________________

529-536

22b. When (on what occasion) is the spraying equipment cleaned?

Every time it is used	01
Every time the pesticide is changed	02
End of the spraying season	03
Not cleaned other than wiping	04
Not cleaned at all	05
DK/CS	06

Others_______________________
(specify) 537-544

22c. Where is the spraying equipment cleaned?

At a flowing water source (tap, river, canal, pump)	1
At a standing water source (lake, pond, well)	2
DK/CS	3
Others________________________	
(specify)	545-546

22d. When was the last time that you disarmed your spraytank for maintenance or to clean the seals?

Less than 3 months ago	1
4 - 6 months ago	2
7 - 12 months ago	3
More than one year ago	4
Never has been opened	5
	547

23a. This last time that you used pesticides, what did you do with the empty pesticide containers?

Washed and re-used	1
Rinsed and sold as scrap	2
Sold as scrap (unrinsed)	3
Broke/burned the container	4
Buried the container	5
Destroyed by ___________________	6
Thrown away in trash can	7
Thrown away into field/bushes	8
Others (specify)	9
	548-553

If "1" coded, ask:
23b. What did you re-use it for?

For storing/carrying _______________(specify item) 554-557

23c. What was the container made of?

Plastic	1	
Glass	2	
Metal	3	
DK/CS	4	558-559

23d. What was its size/capacity?

Less than one litre	1
1.1 - 5 litres	2
5.1 + litres	3
DK/CS	4
	560-561

23e. Where and how did you wash the container?

Where		How		
In flowing water source	1	With water alone	1	564-565
Still water source	2	With water and soap	2	
DK/CS	3	With mud/ash	3	
Others (specify)	4	DK/CS	4	
562-563		Others	5	

24a. After you finished last spraying operation, what did you do thereafter?

Went home directly	1	Go to Q 24b	566
Went to	2		

____

If respondent did not go home, trace his movements till he got home by asking
"and then went to" ____
"and then went to" ____
567-570

If "2" coded, note the response verbatim and post code in the boxes below:

Box 1		Box 2	
Continued to work at farm	1	Activities related to cleaning of equipment	1
Went elsewhere	2	Activities related to personal cleaning/ changing	2
Others (specify) ____________	3	Activities related to other farming work	3
		Others (specify)____________	4

575-578

24a (1). *ASK*: In your estimation, how much time had gone by between the time you finished spraying and till you got home?

Less than 30 minutes	1
30 - 60 minutes	2
1 - 1.5 hours	3
1.5 - 2 hours	4
Over 2 hours	5

579

24b. What did you do after you got home? What else?

	Washed with:		Rank in order of mention
	Water alone	Water and soap	
Had a full body wash	1	2	________
Washed upper body	1	2	________
Washed face and hands	1	2	________
Washed hands only	1	2	________
Washed feet, legs	1	2	________
Changed clothes	X	X	________
Had a meal	X	X	________
Slept	X	X	________
Did nothing	X	X	________
Others (specify)	X	X	________

580-584 585-596

24c. In your estimation, how much time had gone by between the time that you finished spraying and the time that you had a wash (*ask as per answer given above*)

Less than 30 minutes	1
30 minutes - 1 hour	2
1 - 2 hours	3
More than 2 hours	4

597

25a. What did you do with the clothes you had worn while spraying?

Kept them away for next use	1
Rinsed them or had them rinsed in water alone	2
Washed then or had them washed with soap/detergent	3
Other response (*note details*)	4

598

25b. The clothes that you use while spraying pesticides, how often are they washed?

Every time they are used	1	
Once in two-three uses	2	
Once in four-five uses	3	
Less often than that	4	601

26. How long after spraying do you wait before:

Delete as appropriate

You allow a person to
enter the field: _______minutes/hours/days 602-603; 604
You harvest the crop: ______minutes/hours/days 605-606; 607

27a. In your opinion, which is the <u>most toxic</u> pesticide you have ever used? 608-609
Name of the pesticide: ____________________________________

27b. If you were using this pesticide, how would you mix this pesticide?

Use bare hands, without stick	1	
Use gloved hands, without stick	2	
Use bare hands, with stick	3	
Use gloved hands, with stick	4	
Use knapsack handle	5	
Shaking the knapsack/ tank	6	
Filling the tank with water and concentrate alternatively	7	610

27c. If you were using this (*name most toxic pesticide ever used*), what precautions would you take?

Note in col. 27c: DO NOT PROMPT

27d. If a person had no constraints at all, what precautions should he ideally take while spraying?

Note in col. 27d; DO NOT PROMPT

Would/ should:	27c	27d
Wear full shirt	01	01
Wear full trousers	02	02
Wear cloth mask on nose and mouth	03	03
Wear plastic face visor	04	04
Wear hat/cap	05	05
Wear gloves for mixing	06	06
Wear cloth apron - front only	07	07
Wear cloth apron - front and back	08	08
Wear plastic apron - front only	09	09
Wear plastic apron - front and back	10	10
Wear plastic shoulder cover	11	11
Spray along wind direction	12	12
Not smoke while spraying	13	13
Not spray continuously	14	14
DK/Cs	15	15
Others (specify)		
	611-622	623-634

Compare answers in 27d with answers in 20a. Note items coded in 27d but not coded in Q20a. For each of those, ask below:

27e. You have mentioned some measures as being ideal which you do not follow. Why do you not ______________________________

Report for all items not followed.

Take codes from Col.27d and note for each column below

Reason	635-636	647-648	661-662	675-676	689-690
I do not feel ill	01	01	01	01	01
It makes no difference	02	02	02	02	02
Impossible to use	03	03	03	03	03
Too cumbersome	04	04	04	04	04

Pesticides no longer have power	05	05	05	05	05
Other precautions related to food/drink taken	06	06	06	06	06
Accustomed to pesticides by now	07	07	07	07	07
People laugh	08	08	08	08	08
Cannot afford to buy	09	09	09	09	09
No reason	10	10	10	10	10
Not necessary	11	11	11	11	11
DK/CS	12	12	12	12	12
Others (specify)	____	____	____	____	____
	637-646	649-660	663-674	677-688	691-700

28a. Overall, what would you say are the main problems associated with the spraying process? *DO NOT PROMPT*

No major problems	1
Weight of the equipment	2
Health problems arising from spraying	3
Specify:____________________	705-712
Poor remuneration	4
DK/CS	5
Others (specify)	6
	701-704

If "3" not mentioned spontaneously, ask:

28b. Overall, would you say there is any health risk associated with the spraying job?

Yes, definitely	1	Go to Q 29a
Yes, sometimes	2	Go to Q 29b
No	3	CONTINUE
		713

If "3" coded in Q 28b, ask:

28c. Why do you believe there is no health risk associated with the spraying job? *Code and go to Q 30a.*

Pesticides are not as powerful/ poisonous as they used to be	01
I follow all safety precautions	02

We are now accustomed to spraying pesticides	03	
DK/CS	04	714-719
Others (specify)		

If "3" coded in Q 28a or "1" or "2" coded in Q 28b, ask:

29a. What are the health risks that a person faces?

____ 720-725

29b. Who faces this health risk to the greatest extent? *Code under "main".* Who else is exposed to this risk? *Code under "others".*

	Main	Others
Only the person who is spraying and no one else	1	1
Everyone involved in the field at that time	2	2
All agricultural workers	3	3
Those who consume the produce on which pesticides have been sprayed	4	4
Others	5	5
	716	727-732

29c. At what stage do people jeopardise their health --- put themselves at a health risk?

After a number of years of spraying	1
Each time when a person sprays	2
Only when spraying powerful pesticides	3
Each time a person does not take proper safety precautions	4
DK/CS	5
Others	6
733	

ASK ALL

30a. Have you yourself ever experienced any problems due to pesticide spraying?

Yes	1	
No	2	
DK	3	734

If "Yes" coded in 30a, ask:
30b. What was the problem you have mainly experienced? *Single code under Col. B*
30c. What others ? *Code under Col.c*
30d. The last time that you sprayed _________days ago, did you experience any health problems? *Code under Col.d.*

	Col.b	Col.c	Col.d
No/ None	1	1	1
Giddiness	2	2	2
Nausea	3	3	3
Headache	4	4	4
Fever	5	5	5
Vomiting	6	6	6
Bodyache	7	7	7
Boils/rashes	8	8	8
Burns	9	9	9
Skin problems	10	10	10
Vision problems	11	11	11
Blurred vision	12	12	12
Shaking/trembling	13	13	13
Convulsions	14	14	14
Fainting	15	15	15
Others	16	16	16
	735-741	742-749	750-759

Ask all
31a. When was the last time you had health problem caused by pesticides?

___________________________ (MONTH AND YEAR) 760-761; 762-763

Ask all those who have coded anything under column (D) of Q.30d
31b. When you faced this health problem this time, after spraying, what did you do about it?

Nothing	1			
Took rest at home	2			
Nothing	3			
Induced vomiting at home	4			
Went to a doctor	5	Private : 1	Government : 2	766
Went to a hospital	6	Private : 1	Government : 2	767

Went to a village practitioner	7
Others (SPECIFY)	

764-765

If "5 or 6" coded ask:

31c. What did the doctor/person at the hospital (clinic) prescribe or give?

Medicines	01	
Injections	02	
Intravenous drip	03	
Stomach wash	04	
DK/CS	05	768-773
Others (specify)__________________		

31d. What was the total expenditure incurred on this treatment, in terms of: (*Write as specified : If nil, note "O"*)

	Rs	
Fees paid to the doctor	__________________	774-778
Travel costs	__________________	779-782
Purchase of medicine/injection	__________________	783-787
Purchase of special food etc.	__________________	788-791
DK/CS	__________________	

31e. How long did you rest?

________________________(days/hours/minutes)

delete as appropriate 792-793; 794

31f. For how long did the problem continue?

________________________(days/hours/minutes)

delete as appropriate 795-796; 797

I now have some other questions related to farming.

32a. Are all insects harmful or are there any that are beneficial?

All are harmful	1	
Some are beneficial	2	798-800
DK/CS	3	
Others (SPECIFY)		

____

If "2" coded, SHOW CARD

32b. Can you tell me which of these are beneficial?

Lady beetle (adult + larvae)	01
Skyplird larva	02
Laceworm larva	03
Others (specify)____________	04
DK/CS	05
	805-812

33. *SHOW CARD*

There are a set of statements here that I will read out to you. As I read out each statement, please tell me the extent to which you agree or disagree with it, i.e. do you agree strongly, neither agree nor disagree, strongly disagree.

S. no.	Statements	Agree	Neither agree nor disagree	Disagree
1	*Pesticides nowadays are too weak to be poisonous for humans	1	2	3
2	*When I spray I feel like a general who fights enemies	1	2	3
3	*These masks and plastic covers while spraying are unmanly	1	2	3
4	*The only essential precaution is that spraying should be along the wind direction	1	2	3
5	*Man has to take risk. Spraying is my risk. I do not mind	1	2	3

6	*A person who sprays pesticides definitely harms his health in the long run, no matter how careful he is.	1	2	3
7	*Spray persons have become immune to the effects of pesticides	1	2	3
8	*I am convinced that today we treat plants too violently. If we treated them more tenderly they would grow better	1	2	3

813-820

COMMUNICATION DATA

1. Over the last three-four years, have you ever received any information which gave advice on... ?

a) How to use pesticide in a safe way? YES 1 NO 2 821
b) How to use pesticide in an effective way? YES 1 NO 2 822

If coded "2" both in Q.1(a) and Q.1(b). TERMINATE

2. *Ask if coded "1" in Q.1(a) else Go to Q.3*
What information did you receive on the safe usage of pesticides?
____ 823-828

3. *Ask if coded "1" in Q.1(b) else Go to Q.4*
What information did you receive on the effective usage of pesticides?
____ 829-834

4(a). Who told you or from where did you receive the information on the subject of safety and/or effectiveness in the use of pesticides ?

For each item coded under Q.4a, ask Q.4b and Q.4c and post code on the basis of code-list mentioned below the table

4(b). When did you hear/see or receive this message?

4(c). Do you recall the company which provided the farmers in your village with the information on the safe and effective usage of pesticides?

Source of information	Q.4a	Q.4b	Q.4c	Name of the company (if coded "2" in Q.4c)
Poster	01			
Calendar	02			
Diary/book	03			
Talk + slide show	04			
Talk + flip chart	05			
Film / video shows	06			
Demonstration plots	07			
Child learnt at school	08			
Sprayer maintenance camp	09			
Shopkeeper/ Dealer	10			
Agricultural worker	11			
Others (SPECIFY)	12			

Q.4(b) 901-912 Q.4(c) 835-900

This year	1	CIBA/NOVARTIS/ Hindustan Ciba-Geigy	01
Last year	2	Any other company (Specify_______)	02
2-3 years ago	3	DK/CS	88
3-4 years ago	4		
Others (___)	5		
DK/CS	6		

913-914

CLASSIFICATION DATA: RESPONDENTS

a) Age: ________________ years

915-916

b) Education:

Never been to school	1
Schooled for less than 8 years	2
Schooled for 8-13 years	3
Attended college but non-graduate	4
Graduate	5
Post graduate	6

917

IF 6 OR 7, VERIFY IF DEGREE RELATED TO AGRICULTURE:

Yes: 1 No: 2

b) Occupation : ______________________________________

919-920

FAMILY/HOUSEHOLD DATA

c) Family religion: ______________________________________

921

d) Total annual (household income) : Rs.________________

922-927

f) Farm and household goods	Yes	No	DK
1. Farm durables owned			
Ox-drawn plough	1	2	3
Thresher	1	2	3
Knapsack Sprayer	1	2	3
Tractor	1	2	3
Diesel/ electric pump	1	2	3
Others			
2. Does farm have irrigation facilities?	1	2	3

928-939

3. Durables owned in the house

Furniture	1	2	3
Radio	1	2	3
Bicycle	1	2	3
Refrigerator	1	2	3
TV Black & White	1	2	3
TV Colour	1	2	3
Others (SPECIFY)			
None of these			

940-942

g) Family size (Total number of people living in the house and sharing a kitchen)

Total
Men
Women
Children (<18 years)

h) Are any of the other family members involved in spraying?

Yes	1
No	2

IF YES, NOTE SEX AND AGE OF SUCH MEMBERS

Name	M	F	Age
	1	2	
	1	2	
	1	2	
	1	2	
	1	2	

THANK AND TERMINATE

Index